Lecture Notes
in Control and Information Sciences 419

Editors: M. Thoma, F. Allgöwer, M. Morari

Ihab Samy and Da-Wei Gu

Fault Detection and Flight Data Measurement

Demonstrated on Unmanned Air Vehicles Using Neural Networks

Springer

Authors

Dr. Ihab Samy
TRW Ltd.
Stratford Road
Shirley
Solihull B90 4AX
UK
Email: isar1@le.ac.uk

Professor Da-Wei Gu
University of Leicester
Department of Engineering
University Road
Leicester LE1 7RH
UK
Email: dag@le.ac.uk

ISBN 978-3-642-24051-5 e-ISBN 978-3-642-24052-2

DOI 10.1007/978-3-642-24052-2

Lecture Notes in Control and Information Sciences ISSN 0170-8643

Library of Congress Control Number: 2011937286

Typeset & Cover Design: Scientific Publishing Services Pvt. Ltd., Chennai, India.

Printed on acid-free paper

9 8 7 6 5 4 3 2 1

springer.com

Dedicated to my parents;

Effat and Samy Abou Rayan

Ihab Samy Abou Rayan was born in Alexandria, Egypt in 1983. He received a first class MEng degree in Electrical and Electronics Engineering from the University of Leicester, UK. In 2005 he joined the Control and Instrumentation Group at the University of Leicester, and in 2009 received his PhD title. He has held two post doctoral positions at the University of Leicester and Cranfield University, UK. The latter involved work alongside several international companies including: Boeing, Rolls Royce, BAE Systems and Thales. He is currently a Senior Control Engineer at TRW Ltd, UK.

Preface

This book is essentially the first author's PhD thesis, which was successfully defended at the University of Leicester in 2009. It explores the feasibility of two technologies in reducing cost and weight of air vehicles. The first is a fault detection and isolation scheme, which uses neural networks to diagnose faults in sensors. The second is a flush air data sensing (FADS) system, which uses pressure orifices on a wing's leading edge to estimate air data such as; airspeed angle of attack and sideslip.

Fault detection and isolation (FDI) can be traced back to the time before the 1940's when industry did not rely so much on highly mechanised processes and it was sufficient to only fix something when it was truly broken. With the development of more complex systems and the introduction of just in time manufacturing, practitioners moved to preventative measures and maintenance began to be performed on a scheduled interval basis. This approach, however suffered great disadvantages, as normal operations were halted at pre-defined fixed intervals regardless of a fault being present or not. This meant that profitable production time was unnecessarily lost. Furthermore, faults occurring between the fixed intervals were undetected and could prove catastrophic, resulting in unscheduled maintenance downtime.

For these reasons, the concept of condition monitoring was introduced, in which the health of a system was monitored on a continuous basis in order to improve reliability and availability. In general, these approaches monitor health in real time with the aim of reducing fault detection time, the number of false alarms, the number of undetected faults and unscheduled maintenance downtime. In this way if some part of the system, e.g. a sensor or actuator fails to perform as expected, this can be detected and acted upon so that the system is still safe to operate within agreed industry standards. Because of the competitive market, there are many terms used for condition monitoring systems, e.g. Integrated Vehicle Health Management (IVHM), Integrated Systems Health Mangement (ISHM), Engine Health Management (EHM) and Health Usage Management System (HUMS).

One way to understanding FDI schemes, is to consider them as forming a building block of a condition monitoring system with other building blocks including: sensors, actuators, communication links, ground base equipment etc. As such we can assume that FDI is the means with which fault diagnosis is performed. With this in mind, let us now consider the different approaches possible to detecting and isolating faults.

Traditionally, FDI methods relied on hardware (also referred to as physical) redundancy, where fault detection is based on a voting scheme comparing the same measurement type from redundant hardware (e.g. sensors). Another traditional approach is based on setting pre-defined (generally defined by the Original Equipment Manufacturer) thresholds on a chosen parameter (e.g. temperature). Hardware redundancy and threshold-based techniques are simple to use, which could be the reason for their popularity as the less mathematically oriented a method is, the more appealing to industry it becomes. In academia, researchers have found that both methods suffer great disadvantages such as setting thresholds at high levels to avoid false alarms caused by measurement noise. Furthermore in closed loop control systems, the control laws tend to dampen the effects of faults and so simply checking the size of the output signals does not give a reliable insight into overall system health. In fact, shortly before the catastrophic disaster of the Challenger Space Shuttle in 1986, the FDI scheme of the main engine was based on thresholds and it was noted that advanced detection systems could have prevented the crash. Other FDI techniques, include; frequency analysis and expert systems (e.g. case based reasoning, and if-then rule logic). The different methods are briefly outlined in this book.

Over the years, there have been a set of methods quickly emerging in the literature which rely on mathematical models in place of redundant hardware. The techniques are generally referred to as model-based analytical redundancy. Examples include, observer based methods, parameters estimation, parity space and many more. In theory, analytical redundancy should perform as well as hardware redundancy but with reduced cost and weight as redundant hardware is replaced with software models. Model-based FDI systems naturally lead on of from the theories of control systems. Both of them are initially designed using plant models with the desire that they will be robust to modelling errors when applied to the real system. This concept of 'virtual sensors' has been around for over 50 years as depicted in the famous survey paper [1].

Most model-based FDI techniques rely on fixed, linear mathematical models to represent the virtual sensors. While simple to implement, they are generally limited to linear time-invariant (LTI) systems. This reduces the chances of application to the real world. As an alternative, novel approaches include nonlinear, online adaptive schemes where the model is continuously tuned to best fit the time-varying system. A perfect example of this is neural network (NN)-based FDI due to their nonlinear structures and online training capabilities. In this book, the authors aim to outline the advantages (and in some cases disadvantages) of NN-based FDI schemes. More importantly, the FDI scheme is applied to a UAV application. Using analytical redundancy for FDI in UAVs is a direction of development in UAVs where cost and weight reduction is much needed.

While not exhaustive, in this book, the authors aim to introduce in more depth; a literature survey of FDI, demonstration of NN-based FDI schemes for single sensor faults, comparative studies of NN-based FDI to traditional fixed model based approaches such as those using the famous Extended Kalman Filter (EKF), and finally demonstrate how NN-based FDI schemes can detect the more realistic scenario of multiple sensor faults. Future work, to validate the work carried out here, is also noted.

In the second part of this book, a FADS system is designed for a real UAV. With air vehicle manufacturers looking to reduce costs, researchers have examined how to extract air data measurements from an array of pressure orifices, which would be a cheaper alternative to the standard air data boom. Air data booms consist of Pitot-static tubes which measure the airspeed, and mechanical vanes which measure the aircraft aerodynamic orientation (i.e. angle of attack and sideslip). Despite their popularity, air data booms are known to have measurement disadvantages in addition to possible malfunctions: accuracy may be adversely affected by boom bending and vibration, probe size and geometry and by the flow interference due to the probe itself. As a result, in recent years more research has been geared towards finding alternative solutions to air data booms. An example is optical air data systems which measure the atmosphere outside of an air vehicle. However, with the primary goal of most air vehicle manufacturers being the reduction of costs, researchers have found the concept of air data measurements using a matrix of pressure ports to be a cheaper alternative to optical systems and air data booms.

The measurement of flush surface pressures to estimate air data parameters has been known for some time and examples include the FADS system developed and tested on the NASA X-15 hypersonic aircraft [2]. In fact, most aeronautical applications of the FADS system originate from the initial tests carried out by NASA in the early 1980s. From the literature, we will find that few examples have been extended to mini UAVs. This motivated the work carried out here. The FADS system is an invaluable alternative to air data booms especially in mini UAVs. This is because current air data booms can be too heavy and expensive for use on a mini UAV. Additionally, due to the dangerous terrains that they can be exposed to, external instrumentation is best avoided.

Furthermore, most applications use either look up tables or complex mathematical descriptions for modelling the relationships between air data and pressure data. In this book, we study the use of neural networks for modelling these relationships. The resulting FADS system is shown to produce accurate air data estimations but more importantly it reduces instrumentation weight and cost by almost 80% and 97% respectively, when compared to a standard air data boom.

The authors of this book, have collaborated with leading international companies in the field of aerospace and have tried as much as possible to cover a wide range of readers from academia and industry. With the large variety of topics found in this book, it is difficult to define one readership community. However, as an attempt to target the right audience, the authors believe that the research material carried out here would target the following community: advanced control engineers and researchers, condition monitoring engineers and researchers from academia and industry, postgraduate students and flight data engineers.

Acknowledgements

The authors would like to acknowledge the valuable contributions of Professor Ian Postlethwaite from Northumbira University, Mr. John Green from Blue Bear Systems Research (BBSR) Ltd., Dr. James Whidborne of Cranfield University and Dr. Emmanuel Prempain of the University of Leicester. The authors are grateful to the Engineering and Physical Sciences Research Council for financial support.

References

[1] Isermann, R., Balle, P.: Trends in the applications of model-based fault detection and diagnosis of technical processes. Control Engineering Practice 5(5), 709–719 (1997)

[2] Cary, J.P., Keener, E.R.: Flight evaluation of the X-15 Ball-Nose Flow-Direction sensor as an airdata system, NASA TN D-293 (1965)

Contents

Nomenclature

AA-NN	Autoassociative NN
ADC	Analogue to digital converter
AR	Aspect ratio
BBSR	BlueBear Systems Research Ltd
BP	Backpropagation
C_L	Lift coefficient
C_D	Drag coefficient
C_M	Pitching moment coefficient
CB_L	Large constant bias fault
CB_S	Small constant bias fault
CFD	Computational fluid dynamics
CH	Convex hull
Cp	Pressure coefficient
DAQ	Data acquisition
DOS	Dedicated observer scheme
DR	Detectability ratio
DT1	Dynamic test 1
DT2	Dynamic test 2
E1	NN estimation error threshold
E2	NN RMS estimation error threshold
E3	Minimum distance between NN input vector and hidden unit centres
EBPA	Extended error back-propagation algorithm
EKF	Extended Kalman filter
EMRAN	Extended Minimum Resource Allocating Network
FADS	Flush air data sensing

FAA	Federal Aviation Authority
FA	False alarm
FD	Fault detected
FD-FA	Fault detected but false alarms present
FND	Fault not detected
FDI	Fault detection and isolation
FDF	Fault detection filter
FTCS	Fault tolerant control system
GD	Gradient descent
GOS	Generalised observer scheme
GPS	Global positioning system
HA_L	Large hard additive fault
HA_S	Small hard additive fault
INS	Inertial navigation system
KF	Kalman filter
LMS	Least mean square
LUT	Lookup table
M	Pitching moment
MAV	Mini air vehicle
MEE	Mean estimation error
Min	Minimum function
MLP	Multilayer perceptron
MMKF	Multiple model Kalman filtering
MPO	Matrix of pressure orifices
MSE	Mean squared error
MT	Mean detection time
NAS	National Airspace
NN	Neural networks
P1	Pressure port 1
P2	Pressure port 2
P3	Pressure port 3

P4	Pressure port 4
P5	Pressure port 5
RBF	Radial basis function
Re	Reynolds number
RGE	Residual generation and evaluation
RGPE	Residual generation, padding and evaluation
RMS	Root mean square
ΔRMS	Rate of change in RMS
SA_L	Large soft additive fault
SA_S	Small soft additive fault
ST_L	Large step-type fault
ST_S	Small step-type fault
SFDIA	Sensor fault detection, isolation and accommodation
TeD	NN testing data set
TrD	NN training data set
UAV	Unmanned air vehicle
UD	Number of undetected faults
VEE	Variance estimation error
X	Axial force
Z	Normal force
t_{fa}	Total false alarm duration (s)
r_{pr}	Maximum residual magnitude prior to fault detection
r_{af}	Residual magnitude at fault detection
$\bar{r}$	Basic residual
Ω	Residual averaging size (in samples)
r_{kRGE}	Residual using RGE method
r_{kRGE}	Residual using RGPE method
ϖ	Residual weight
p_{pad}	Number of padding points
q-NN	Pitch rate NN
α-NN	Angle of attack NN

a_z-NN	Normal acceleration NN
λ	NN connecting weights
$\boldsymbol{\mu}$	RBF centre
σ	RBF width
$\boldsymbol{\mu}_{\mathbf{rk}}$	Centre of hidden neuron closest to the input vector
ε_{max}	Initial *E3* threshold
ε_{min}	Final *E3* threshold
γ_{dy}	Decay factor for *E3* threshold
e_k	NN estimation error
k_{op}	NN overlap factor
N_{max}	Maximum number of hidden neurons
$\boldsymbol{\theta}$	Vector of NN free parameters
δ	NN learning rate
T_R	Fault ramp duration
t_{fault}	Fault start time
A	Fault magnitude
p	pressure
p_i	Surface pressure
θ_i	Flow incidence angle
λ_i	Angle the normal to the surface at port *i* makes with the longitudinal axis of the nosecap
ϕ_i	Clockwise angle looking aft around the axis of symmetry starting at the bottom of the nosecap
ε	Calibration parameter
E	Instantaneous squared error
P_∞	Freestream static pressure
P_0	Total pressure
α	Angle of attack
β	Sideslip
α_{eff}	Local angle of attack
β_{eff}	Local angle of sideslip

T_{∞}	Freestream temperature
q	Pitch rate
q_{∞}	Freestream dynamic pressure
S	Wing area
b	Wing span
$\bar{\bar{c}}$	Mean aerodynamic chord
m	Mass
x_{cg}	Centre of gravity location (with respect to nose)
I_x	Roll inertia
I_y	Pitch inertia
I_z	Yaw inertia
I_{xz}	Product inertia
u	Axial velocity
w	Normal velocity
v	Lateral velocity
p	Roll rate
r	Yaw rate
q	Pitch rate
Φ	Roll angle
θ	Pitch angle
ψ	Yaw angle
P_n	North position in earth axis
P_e	East Position in earth axis
h	Altitude
ρ	Air density
V_t	Airspeed
z_t	z-axis coordinate of the engine thrust line
τ	Throttle setting
T_X	Thrust
γ	Flight path angle
η	Elevator angle

α_{gust}	Gaussian gust disturbances acting on the angle of attack
β_{gust}	Gaussian gust disturbances acting on the angle of sideslip
a_x	Axial acceleration
a_y	Lateral acceleration
a_z	Normal acceleration
C_{L_α}	Lift coefficient due to angle of attack
C_{L_η}	Lift coefficient due to elevator demand
V_∞	Freestream airspeed
V	Airspeed
ω	System noise
$\mathrm{x_u}$	Axial force due to axial velocity
$\mathrm{x_w}$	Axial force due to normal velocity
$\mathrm{x_q}$	Axial force due to pitch rate
$\mathrm{x_\theta}$	Axial force due to pitch angle
x/c	Normalised wing chord
t/c	Thickness to chord ratio
$\mathbf{v}$	Measurement noise vector
Q	System noise variance
R	Measurement noise covariance matrix
$\hat{\mathbf{x}}_\mathbf{k}^-$	Predicted (a priori) state estimate
$\hat{\mathbf{x}}_\mathbf{k}$	Corrected (a posterior) state estimate
$\mathbf{P_k}$	State estimation error covariance matrix
$\mathbf{K_k}$	Kalman Gain

Subscripts

real	Real sensor measurement
$\hat{}$	Estimate
NN	Neural Network estimate
EKF	Extended Kalman filter estimate
Ideal	Sensor measurement when no faults are present
∞	Freestream

Chapter 1
Introduction

1.1 Research Objectives

Nowadays unmanned air vehicles (UAVs) are used in applications deemed dangerous or unreachable by manned air vehicles. It is estimated that nearly 8000 UAVs (worth $3.9 billion) were produced worldwide between 1994 and 2003 [1]. Their applications have been widespread and included: the agricultural industry, environmental control, mineral exploration, news broadcasting, unexploded mine exploration and other military related applications [2]. Furthermore the design of UAVs has been widespread ranging from simple fixed-wing aircraft [3, 4] to complex quadrotor air vehicles [5, 6]. One of the reasons behind this growing interest, especially in the military industry, is that UAVs are relatively cheap to manufacture in comparison to current fighter aircraft.

UAVs are quite popular in the military industry due to the dangerous terrains experienced. However current trends in UAV design have shown that cheap and low weight UAVs are also likely to be accepted by the civil aviation industry [7]. This book therefore aims to exploit existing manned air vehicle technologies to reduce the instrumentation costs and weight of UAVs. Two technologies are investigated: model-based sensor fault detection, isolation and accommodation (SFDIA) schemes and flush air data sensing (FADS) systems.

Fault detection and isolation (FDI) can be traced back to the era before the 1950's where a fix-it-when-it-broke approach was sufficient, as industry did not rely on highly mechanised processes. With the development of more complex systems and the introduction of just-in time manufacturing mindsets, practitioners found the concept of preventive maintenance to be more efficient, where maintenance was performed on a scheduled interval basis. However this approach suffered great disadvantages, as operations were halted at pre-defined fixed intervals regardless of a fault being present. This meant that profitable production time was lost. Furthermore faults occurring between the fixed intervals were unaccounted for and could prove catastrophic, if undetected, resulting in unscheduled maintenance downtime.

As such, the concept of condition monitoring systems was found to be more appealing where the health of the system was monitored on a continuous basis in order to improve system reliability and availability. In general condition monitoring systems monitor the health of the platform in real time with the aim of reducing fault detection time, false alarm rates, undetected faults and unscheduled maintenance downtime. It must be noted that condition monitoring systems are seen as support systems to already existing safety-critical fault detection systems.

I. Samy and D.-W. Gu: Fault Detection and Flight Data Measurement, LNCIS 419, pp. 1–4.
springerlink.com

In other words if the condition monitoring system fails to perform as expected, the platform will still be safe to operate. This is important if the platform is to meet specific industry safety standards (an example is the stringent airworthiness standard set by EASA or FAA for new aircraft).

There are several commercial names for condition monitoring systems due the highly competitive industry. Examples include: System 1, IVHM, ISHM, HUMS, EHM. However, in general most condition monitoring systems can be broken down to few building blocks. FDI is one of the building blocks of a condition monitoring system and has been the focus of intense research in the academic community over the past 40 years or so. This book considers (including literature surveys and comparison results) FDI with application to UAVs.

Traditionally SFDIA schemes are based on physical redundancy where multiple sensors are used to measure the same flight parameter. Sensor faults are then detected based on a simple voting scheme. Despite its simplicity this approach can be expensive to implement especially if multiple sensors are used on board the air vehicle. As an alternative, research has shown that model-based SFDIA schemes (analytical redundancy) can perform just as well as physical redundancy techniques, but with fewer incurred costs. The concept of 'virtual' sensors has been around for over 40 years [8]. Examples of popular survey papers and books in this field include [8-18]. Most of the work carried out so far has considered fixed, linear model-based approaches, with parameter estimation and observer-based methods being the most popular [8]. While proving to be successful they are generally limited to linear time-invariant (LTI) systems. Novel approaches include nonlinear, online adaptive schemes where the model is continuously tuned to best fit the time-varying system. An example of such methods is neural network (NN)-based SFDIA schemes due to their nonlinear structures and online training capabilities. In the famous survey paper by Isermann and Balle [8], it was noted that NN-based SFDIA schemes were steadily growing in number especially in nonlinear applications. Examples of such methods include [19-28]. However few have been extended to UAV applications. SFDIA via sensor redundancy may not be an option in UAVs (such as mini air vehicles (MAVs)) due to cost, weight and space restrictions. Therefore NN-based SFDIA schemes are an invaluable solution to UAV applications.

The second part of this book investigates the use of FADS systems for air data measurements. One of the most popular instruments used for air data measurements is the air data boom. Air data booms consist of Pitot-static tubes which measure the airspeed, and mechanical vanes which measure the aircraft aerodynamic orientation (i.e. angle of attack and sideslip). Air data booms can be too heavy for small UAVs (as will be discussed in Chapter 7, Conclusions section). Furthermore with the primary goal of most UAV manufacturers being the reduction of costs, researchers found the concept of air data measurements using a matrix of pressure orifices to be a cheaper alternative to air data booms. The concept of FADS systems is not new and has been implemented by several research groups over the past 30 years [29-45]. However, as far as the author is aware, the FADS system has not yet been tested on MAVs. MAVs are found within the spectrum of UAVs and are characterised by their low costs, small size and low weight. As such, air data booms may not be suitable in MAVs and therefore the FADS system is a promising alternative for air data measurements in MAVs. In this book we design and test a FADS system on a MAV (supplied by BlueBear Systems Research (BBSR) Ltd.). Our work is distinct from previous research in that: 1) A FADS system is designed and implemented on the wing of a MAV which flies at speeds as low as Mach 0.07 and 2) an extended minimum resource allocating network (EMRAN) radial basis function (RBF) NN is trained to model the aerodynamic relationships in the FADS system.

1.2 Book Contributions

The main aim of this book is to exploit existing aircraft technologies for the reduction of costs and weight in UAVs. The objectives and contributions of this book can be summarised as follows:

1. To deliver a complete literature survey of fault detection and isolation systems
2. To apply existing aircraft technologies to UAVs. The technologies include NN-based SFDIA schemes and FADS systems.
3. To compare the SFDIA performance of NN-based methods to traditional fixed model-based methods. An extended Kalman filter (EKF) is chosen as a representative of fixed (nonlinear) model-based approaches which rely on a mathematical description of the real system.
4. To design and implement a NN-based SFDIA scheme to detect single and multiple sensor faults in UAVs. A nonlinear UAV model is used as a test bed and the performance of the NN based SFDIA scheme is investigated under different levels of system and measurement noise and different sensor fault types.
5. To improve the robustness and sensitivity of model-based SFDIA schemes to unknown inputs (e.g. measurement noise) and incipient faults (small and slow drifting faults) respectively. A novel residual processing technique referred to as residual padding is proposed. Residual padding aims to reduce the false alarm rates and number of undetected faults in current model-based SFDIA schemes.
6. To investigate the feasibility of using a FADS system on a MAV. As far as the author is aware the FADS system has so far been applied to large, manned air vehicles. Moreover most applications tend to place the FADS system at the nosecap of the aircraft. This may not be an option in MAVs due to the presence of a nose propeller. Alternatively we consider mounting the FADS system on the wing leading edge.
7. Traditional approaches to modelling the relationship between aircraft surface pressure and air data are based on aerodynamic models or lookup tables. In this book a NN is trained to relate the surface pressure to the air data states. The NN-based FADS system results are also compared to a standard lookup table approach.
8. FADS systems are based on pressure measurements from orifices drilled into the aircraft surface. As most MAVs are flown at low altitudes, the orifices are susceptible to blockage from poor weather conditions and atmospheric debris. For this reason we investigate the robustness of the FADS system to faults and propose several methods for fault accommodation purposes.

1.3 Book Structure

This book is organised as follows (see road map, Fig 1.1): In Chapter 2 we discuss the general background of fault detection and identification (FDI) schemes, the terminologies used and the pioneers in this vast field. In Chapter 3 we introduce the concept of FADS systems. In Chapter 4 the NN model used in both the SFDIA scheme and the FADS system is outlined. Chapters 5-7 are the application chapters. Chapters 5 and 6 present the SFDIA tests carried out for single and multiple sensor fault scenarios respectively, while Chapter 7 presents the FADS system designed for the MAV. Finally, Chapter 8 concludes the work in this book and proposes the future work that can help to better understand and extend the work carried out here.

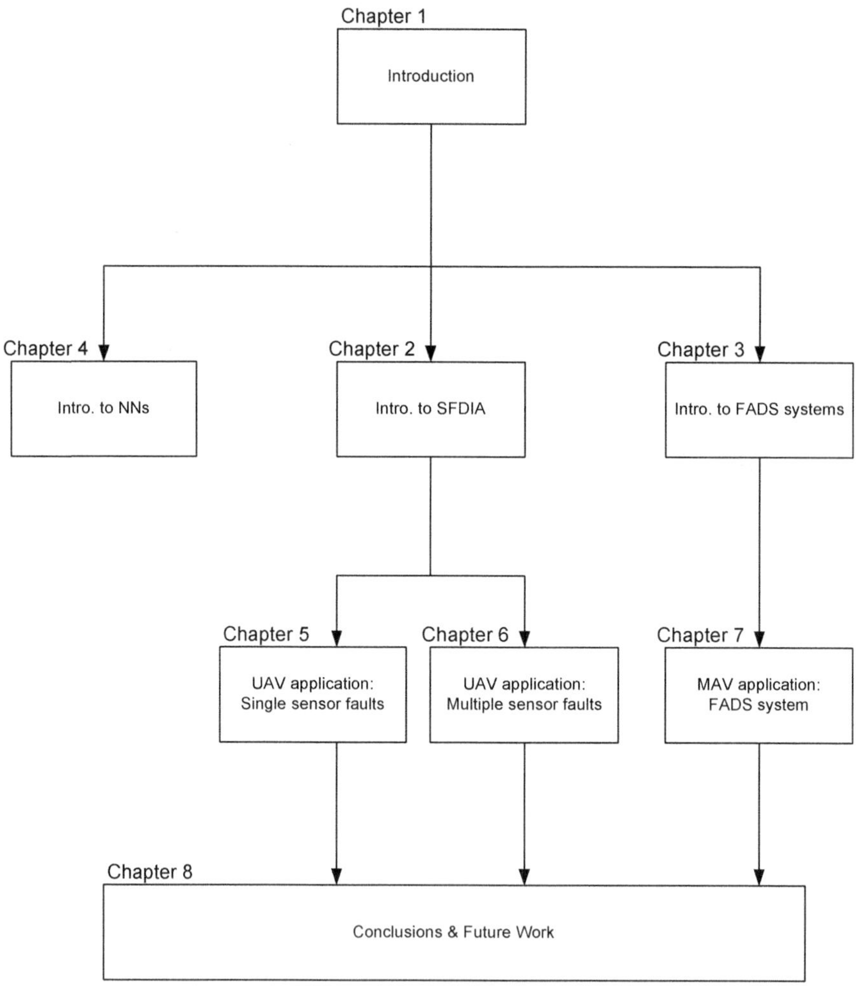

Fig. 1.1 Book road map

Chapter 2
Fault Detection and Isolation (FDI)

Introduction

With the growing use of complex systems, there has been considerable interest in the development of techniques to detect and isolate faults. An undetected fault in a system can have catastrophic effects such as loss of human life, environmental pollution and financial losses. Examples of places where FDI schemes can be useful are hospitals and manufacturing companies. In hospitals, staff would need to be aware of faults in health monitoring equipment (e.g. electrocardiographs) to avoid incorrect patient health diagnosis. On the other hand an undetected fault in a production line can eventually require overall plant shutdown, which can be costly. The literature and effort gone into the field of FDI is overwhelming. It still remains one of the most active areas of research today. Owing to this, one can expect the terminology to be quite misleading as different authors assign different terms to describe similar concepts. To address this problem, in 1997 the IFAC Technical Committee: SAFEPROCESS (Fault detection, supervision and safety for technical processes) grouped commonly accepted definitions that seem to be consistent throughout the field [8]:

- **Unknown inputs:** These include unmodelled disturbances (system noise), measurement (sensor) noise, modelling uncertainties and system parameter variations [9].
- **Fault:** An unpermitted deviation of at least one characteristic property or parameter of the system from the acceptable/usual condition.
- **Residual:** A fault indicator, based on a deviation between real measurements and model- based estimates.
- **Fault detection:** Determination of the faults present in a system and the time of detection.
- **Fault isolation:** Determination of the location/cause of a fault. Follows fault detection.
- **Fault accommodation:** Means by which system safe operation is maintained in the event of a fault. It follows fault isolation.
- **Analytical redundancy:** Use of two ways to determine a variable, where one way uses a model in analytical form (i.e. a computer program).

In general most FDI methods can be divided into two groups (Fig 2.1) [10]:

- One that makes use of a plant model
- One that does not make use of a plant model

I. Samy and D.-W. Gu: Fault Detection and Flight Data Measurement, LNCIS 419, pp. 5–17.
springerlink.com

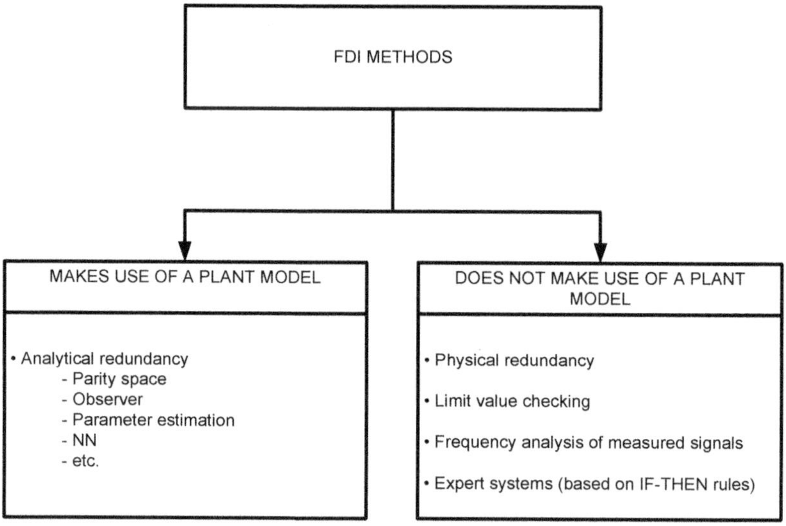

Fig. 2.1 FDI methods

Traditional FDI methods rely on redundant hardware (e.g. sensors) where fault detection is based on some sort of voting scheme. This approach is known as FDI via *physical redundancy*. Examples of such applications can be found in [52, 53].

Another popular approach is based on simple *limit-value checking* of characteristic variables (e.g. temperature) [11]. Limit-value checking remains one of the most widely implemented FDI methods in industry, due to its simplicity [10]. Tolerances (also referred to as thresholds) are either recommended by product manufacturers or defined from experience. Unfortunately, limit-value checking techniques are only reliable if faults are large or long-lasting. This is because thresholds are set at high levels to avoid false alarms caused by random system fluctuations. Furthermore in closed loop systems, the control laws tend to dampen the effects of faults and so simply checking the size of the output signals does not give a reliable insight into overall system health. Shortly before the catastrophic disaster of the Challenger Space Shuttle in 1986, the FDI scheme of the main engine was outlined in [54]. It was noted how limit value checking methods were mainly used for engine health monitoring and suggested that advanced failure detection systems are needed [54].

Frequency analysis of measured signals can also give invaluable insight into machine health. These methods are extremely popular if faults cause an increase in machine vibration. The frequency spectrum of these vibrations can therefore be used for FDI purposes. A good survey paper discussing vibration-based FDI methods can be found in [55].

The use of *expert systems* (knowledge-based methods) is another popular FDI approach. Expert systems rely on what are known as heuristic symptoms (e.g. measurable symptoms, machine performance history, etc.) to detect and isolate faults. Fault detection is based on qualitative information which can be provided from knowledge of the system health history (e.g. former faults, maintenance performed etc.), or from human observations (e.g. smell, sound etc.). Fault isolation is then based on IF-THEN logic or pattern classification techniques using neural networks [8, 10]. Expert systems have received considerable attention over the years and [5] pointed out that analytical redundancy can be combined with expert systems for a more informative FDI scheme. The reader is referred to the chapter of Tzafestas in [9] for an introduction to expert systems, and [56, 57] for examples of such systems.

Over the years, strong interest in control theory has brought about powerful techniques in mathematical modelling which have been made feasible due to the progress of modern computer technology [12]. Consequently researchers found the use of such models, as direct replacements of redundant hardware (i.e. physical redundancy), a cost and weight effective approach to FDI. Moreover these models are capable of estimating states that are often non-measurable which can give an invaluable insight into plant operation and simplify the fault isolation process. Collectively these approaches are known as FDI via *analytical redundancy* or *model-based* FDI. Over the past 30-40 years, FDI via analytical redundancy has experienced a wide variety of theoretical contributions and applications. However the great variety of proposed methods can be brought down to a few well known techniques. Some of these will be discussed in section 2.1. Popular survey papers include [8, 10, 12-15, 58, 59].

Model-based FDI systems naturally lead on from the theories of control systems [60]. Both of them are initially designed using plant models with the desire that they will be robust to modelling errors when applied to the real system. An inadequate control law can result in instability in the real system while an inadequate FDI scheme can result in high false alarm rates and undetected faults. The combination of a FDI scheme and a control system is known as a fault tolerant control system (FTCS). FTCS can benefit from the ability to compensate for faults (detected and isolated by the FDI scheme) while maintaining satisfactory system performance [61].

Despite the extensive research gone into model-based FDI methods, one will find that unlike control theory, they have not been utilised much in industry. Blanke and Patton suggest that this is due to the scarcity of realistic examples [62]. Survey papers and books in the field of FDI are widespread but real industrial applications are not. One reason may be that despite the attractive theory developed over the years, practitioners found that physical redundancy and traditional limit-value checking techniques can deliver satisfactory results with less theoretical effort. For example in spacecraft where production costs are high, the addition of redundant sensors for sensor FDI (SFDI) may not be significant in comparison to the effort required to model such a complex system. However, there are some applications where physical redundancy may not be an option and so model-based FDI schemes become an invaluable alternative. This is true of UAVs which have limited onboard space, weight restrictions and demand low production costs.

2.1 Model-Based FDI

A plant can generally be divided into three subsystems (Fig 2.2); actuators, the process (i.e. components) and sensors. For example in an aircraft, the actuators include the control surfaces (elevators, ailerons, rudders), the process would include the actual airframe, and the sensors would be the instrumentation onboard the aircraft. Chow and Willsky first defined model-based FDI schemes to involving two stages; *residual generation* and *residual evaluation* (Fig 2.2) [63]. A system model is used to generate a residual which is usually a function of the difference between the model estimate and the real measurement. This stage is referred to as residual generation. The FDI decisions are then carried out in the residual evaluation stage. It is important to note that in most cases the method of residual evaluation is greatly dependent on the method of residual generation. This will become clearer when we discuss the different residual generation approaches in sections 2.1.1-2.1.5.

In general, faults in a plant can be divided into three categories:

1) Actuator faults (additive)
2) Process faults (additive or multiplicative)
3) Sensor faults (additive)

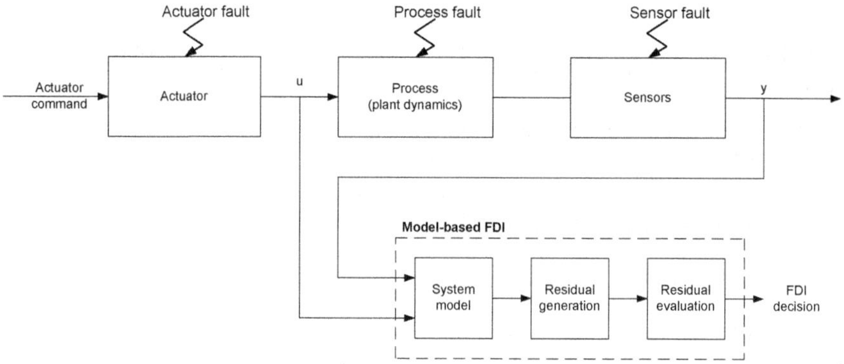

Fig. 2.2 Model-based FDI scheme

Actuator faults can for example cause a malfunction in the engine of an industrial process or a fault in the control surface of an aircraft. They are additive faults in the sense that they influence the system with an additive term. Sensor faults are also additive faults but influence the instrumentation (sensors) of the system. For example they may include sensor biases, total sensor failures or sensor drifts. On the other hand, process faults can be either additive or multiplicative. Parametric faults are examples of multiplicative process faults, i.e. the faults influence the plant output by the product of another variable [16]. They result in changes in the plant parameters. For example the deterioration of an aircraft's airframe would be considered a parametric fault. Process faults can also be additive. They include unmeasured disturbances acting on the plant which are normally zero e.g. plant leaks [10]. It is important to classify the different fault categories. This is because different FDI schemes are better suited for different types of faults. So for example parameter estimation methods (section 2.1.4) are best suited for parametric faults while observer-based schemes (section 2.1.2) are more popularly used to detect actuator and sensor faults. In fact each fault category is a research field in its own right; actuator FDI (AFDI), component FDI (CFDI) and sensor FDI (SFDI) [9].

In addition to the different fault categories, faults can also be; abrupt (quick-varying) or incipient (slow-varying) [64]. Abrupt faults cause a sudden change in the nominal (fault-free) behaviour of the system while incipient faults have drift-type effects. For example constant bias sensor faults can be considered as abrupt faults whereas soft additive faults have a much slower effect on the sensor measurements and are therefore referred to as incipient faults. Incipient faults are usually caused by temperature drifts, calibration problems or worn equipment. In the short term they may not cause significant performance degradation. As such there are generally no tight constraints on their fault detection time. However they could prove catastrophic if left undetected for too long [9].

From the literature one will realise that there are a wide variety of model-based FDI methods. However most of them can be brought down to a few well-known techniques [8, 12]:

1) Parity space
2) Observer-based
3) Fault detection filter (FDF)
4) Parameter estimation
5) Neural networks

There are of course many other model-based FDI methods which deserve similar, if not more, attention. Examples include; FDI via eigenstructure assignment first proposed by [65], and FDI via sliding mode observers proposed in [66]. Fortunately the literature in this field is widespread and the reader is referred to [11, 17] for a general introduction.

2.1.1 Parity Space

A popular class of model-based FDI schemes is the parity space approach. Some of the first pioneers in FDI via parity equations were Chow and Willsky [63], and Lou *et* al. [67]. Other contributors include [18, 68-74]. Parity equations are residual generators which generate the residuals by direct manipulation of the plant observables (i.e. measured inputs and outputs). In this approach, a number of plant observables are sampled from previous time instants to a current time instant. The residual generated is then calculated as a function of these sampled measurements and a user-defined matrix. By suitably designing this matrix, the residual can equal zero when no faults are present and nonzero when a fault is present.

2.1.2 Observer-Based

Observer-based methods are one of the most popular approaches to model-based FDI [8]. Observers are often used in control systems to estimate non-measurable states (required in e.g. health monitoring systems or control laws). An observer is essentially a system model, and can fall into one of two categories: a Luenberger observer (used in a deterministic setting) or a Kalman filter (used in a stochastic setting) which were originally proposed in [75] and [76] respectively. The estimation errors of the Luenberger observer (often simply referred to as observer) or the innovation sequence of the Kalman filter can be used as residuals for FDI purposes. Consider the following state-space system with additive faults included:

$$\dot{x}(t) = Ax(t) + Bu(t) + Lf_L(t) \tag{2.1}$$

$$y(t) = Cx(t) + Mf_M(t) \tag{2.2}$$

where $x \in \mathcal{R}^n, y \in \mathcal{R}^p, u \in \mathcal{R}^m$, input and output additive faults are represented by the fault vectors f_L and f_M respectively, and L and M are distribution matrices for the input and output faults respectively. A Luenberger observer for the system in (2.1)-(2.2) can be designed as in Fig 2.3 and defined as:

$$\dot{\hat{x}}(t) = A\hat{x}(t) + Bu(t) + H\big(y(t) - \hat{y}(t)\big) \tag{2.3}$$

$$\hat{y}(t) = C\hat{x}(t) \tag{2.4}$$

where $\hat{y}$ and $\hat{x}$ are the observer state and output estimates respectively, H is a user-defined feedback matrix and the following are the state and output estimation errors:

$$\varepsilon = x - \hat{x} \tag{2.5}$$

$$e_y = y - \hat{y} \tag{2.6}$$

System states are often non-measurable and therefore the output estimation error e_y is instead used as the residual. Substituting (2.1)-(2.4) into (2.5)-(2.6) the following can be defined:

$$\dot{\varepsilon} = (A - HC)\varepsilon + Lf_L - HMf_M \tag{2.7}$$

$$e_y = C\varepsilon + Mf_M \tag{2.8}$$

The residual in (2.8) is a function of the additive faults (f_L and f_M) and therefore, with the proper design of H, the residual can be made non-zero only in the event of a fault. It is important to note that observers designed for control systems have different purposes than for FDI schemes. In FDI applications, a non-zero e_y triggers a fault alarm which fulfils the purpose of the observer, while in control systems this would not be ideal as accurate state estimations are consistently needed in the feedback control laws.

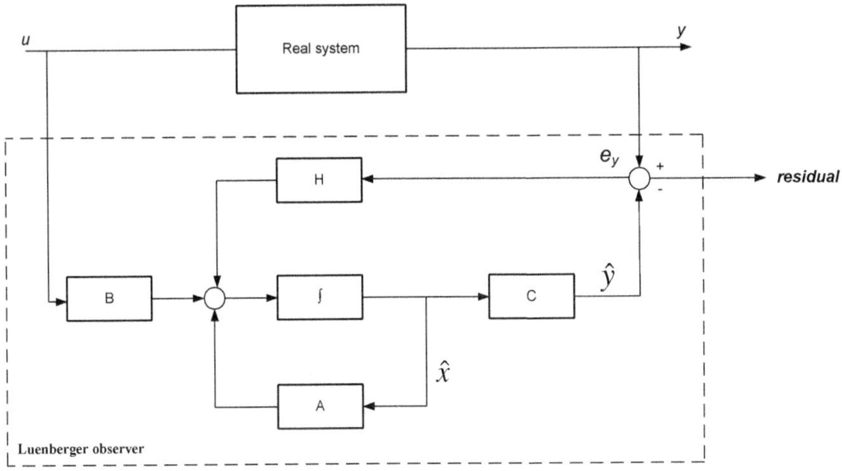

Fig. 2.3 Luenberger observer

A potential problem with the observer in Fig 2.3 is that a single fault seen in the output measurement vector *y*, can contaminate all observer estimates through the feedback matrix *H*. This therefore complicates the fault isolation process. To tackle this problem, several well-known observer-based FDI schemes have been proposed:

a) **Simplified instrument FDI:** This was originally proposed in [77] for FDI of sensor faults. One observer driven by only one sensor, y_i, is used to estimate all system outputs. Therefore if y_i is faulty then all residuals will be non-zero. On the other hand if the other sensors are faulty (i.e. any sensor except y_i) then only the corresponding residual will be non-zero. In principle this approach can detect simultaneous faults (i.e. where more than one sensor can fail at a time) as long as the faulty sensor is not y_i. Note that the observer used can be a Luenberger observer, a Kalman filter or an unknown input observer (UIO) [11]. In [78], Clark shows an example which makes use of a Kalman filter in the event of unmodelled disturbances.

b) **Dedicated observer scheme (DOS):** This was again suggested by Clark in [79] for FDI of simultaneous sensor faults. It is an extension to a) in that a *bank of observers* (also referred to as an *observer scheme*) are used instead of one observer. Each observer is dedicated to only one sensor, i.e. driven by only one sensor. In the event of a fault the corresponding observer will produce inaccurate estimates and therefore all of its residuals will be non-zero. DOSs can also be used for FDI of actuator faults, if the faults are associated with direct changes in the input signal *u(t)* [9]. This time each observer is driven by all sensor measurements but only one input measurement, u_i.

c) **Generalised observer scheme (GOS):** In [80], Frank suggested the use of a GOS for FDI of sensor faults. As in the DOS, this method makes use of a bank of observers. However this time each observer is driven by all sensor measurements except for one. In principle this allows the detection and isolation of a single fault (i.e. only one sensor can fail at a time) but with improved robustness to unknown inputs.

d) **Hypobook testing:** These methods were popular in the 1970's and 1980's [13]. Mehra and Peschon [81] suggested that the statistical properties (e.g. mean, variance) of the Kalman filter innovation sequence (difference between the measured system outputs and the Kalman filter estimates) change when a fault is present and so by observing this change, one can in principle detect the fault. However this method lacked the ability to isolate the fault and was shown simply to be an alarm system [23]. In [82] and [83] Willsky and Jones suggested an extension to this approach where the innovation sequence is compared to pre-determined fault hypotheses, i.e. fault isolation is made possible by using *a priori* knowledge of the different effects the failures have on the innovation sequence. Another popular method was also published in [84] and [85]. It was referred to as the multiple model Kalman filtering (MMKF) approach. This time a bank of Kalman filters are used where each filter represents one specific failure mode and faults are isolated using a multiple hypotheses test. Despite their popularity these methods were deemed quite computationally complex. Moreover if the real system fails in a way which is different than the pre-determined failure models, the fault will pass by undetected. As a result some of these techniques have lost much of their popularity over the years [10]. However, excellent research is still being carried out [86-89].

Observer-based methods can be of full order (see e.g. [90]) or reduced order (see e.g.[79]) and can be designed for application in nonlinear systems where for example an extended Kalman filter (EKF) can be used instead of the linear Kalman filter, see e.g. [91]. Another famous observer-based method, originally proposed by Edwards and Spurgeon, is the sliding mode observer [66]. For a wider introduction to observer-based FDI schemes the reader is referred to [11] and [17].

2.1.3 Fault Detection Filter

Originally proposed in Beard [77] and redefined in Jones [78], this technique heavily relies on the design of a feedback matrix H (Fig 2.3) so that each fault causes the residual to lie in a specific direction. The faults are then isolated by observing the direction in which the residual propagates. The FDF is a powerful and popular method for FDI schemes and in some publications is included under observer-based methods as it also relies on an observer as in Fig 2.3.The reader is referred to [11], [13], [17] [92, 93], for a wider introduction to FDFs.

2.1.4 Parameter Estimation

Parameter estimation methods are a popular approach to FDI of parametric faults [14]. In contrast to observer-based methods, they assume that the system model is unknown. The input/output measured signals are instead used to estimate the process model and its associated parameters. The benefit from this is that the parameter estimates are in fact related to what are known as *process coefficients* which are in turn related to the faults. Examples of process coefficients include; stiffness, length, mass, resistance, speed etc. Once estimated, process coefficients are then compared to pre-defined reference values (e.g. a fixed threshold) for fault detection purposes. Fault isolation is then implemented based on the knowledge of the relationship between the faults and the process coefficients. A simple example is a resistor in a circuit, which increases in resistance when faulty. Therefore if we can estimate and monitor this resistance, we can detect the faults in the resistor. For examples of FDI via parameter estimation and a general discussion the reader is referred to [58], [9] and [94], [18].

2.1.5 Neural Networks

A NN-based FDI scheme uses a NN to replace traditional models which rely on a detailed mathematical description of the system (such as observers or Kalman filters, section 2.1.2). They are nonparametric models in the sense that they build their structure purely from training data. In the early stages of FDI, one will find that most of the literature does not propose NN-based solutions to FDI [9, 10, 12, 13]. This is despite the fact that NNs have been around since 1943 when the first model was proposed by McCulloch and Pitts [95]. One reason for this lack of interest is the bad media coverage that NNs received after Minsky's and Papert's publication on the limitations of NNs [96]. At that time the NN was seen as only being applicable to pattern classification problems which must be linearly separable (e.g. the popular XOR mapping problem could not be solved, see [97]). Moreover multilayered NNs were shown to suffer from what is known as the credit assignment problem [98]. To explain this further, consider a NN with only one hidden layer and one output layer. The desired NN output is usually known and therefore parameters in the output layer can be tuned accordingly. On the other hand the desired output of the hidden layer is unknown and so it can be difficult to correctly tune the hidden layer parameters. This is known as the credit assignment problem and was originally suggested by Minsky in [98]. However to address this issue Rumelhardt & McClelland published their famous book which proposed the well known backpropagation (BP) training algorithm for training multilayered perceptrons (MLPs) [97]. The MLP NN trained via the BP algorithm remains one of the most popular NN architectures today [97]. Other architectures also include the radial basis function (RBF) NN which will be discussed in Chapter 4. Since then NNs have slowly found their way into a wide variety of engineering applications and eventually into FDI applications. In the late 1980's and early 1990's they became quite popular for FDI in chemical processes [100-102] and in the famous survey paper by Isermann and Balle, it was noted that

NN-based FDI schemes were gradually increasing [8]. NN models can be used as pattern classifiers in FDI applications where each fault type has a different symptom. Alternatively NN models can be used in the same layout as other model-based FDI schemes except that the NN would replace the corresponding model. For example in the GOS (section 2.1.2), a NN model can replace each observer (see e.g. [19]). Another example is in parameter estimation FDI schemes (section 2.1.4). In this case, a NN can be trained to estimate the chosen parameters (see e.g. [20]). For more examples of NN-based FDI schemes and for a wider introduction to NNs, the reader is referred to [19-28] and [97, 103, 104] respectively.

2.2 Performance Criteria

In section 2.1 we discussed some of the popular model-based FDI methods. Choosing the 'best' method is not an easy task and generally depends on the application (e.g. parameter estimation methods are better suited to detect parametric faults). Furthermore assessing the performance of each method requires a universally accepted benchmark which can be difficult to define. However there are a set of performance criteria that have repeatedly emerged in the literature which allow us to compare the different FDI schemes:

1) Fault detection time
2) False alarm rate
3) Number of undetected faults
4) Ability to isolate the fault

A low fault detection time is desirable if we are to avoid any permanent damage in the system. However its level of importance depends on the application. For example in aerospace applications, faulty sensor measurements used in the control laws can potentially result in flight instability and therefore it is desirable to detect the fault as early as possible. The drawback of designing the FDI method based purely on its fault detection time is that other performance criteria (particularly the false alarm rate) are often compromised. For example selecting a low residual threshold can reduce the fault detection times but it can also increase the false alarm rate.

The robustness of a model-based FDI scheme to unknown inputs can be generally judged based on its false alarm rate. For example observer-based methods which rely on a linearised state space description of the plant will be prone to linearization errors. Similarly a Kalman filter which assumes stationary (i.e. fixed statistics) noise signals will be prone to modelling errors when applied to a real system where noise is generally non-stationary. These modelling errors can cause the residuals to be non-zero even when a fault is not present resulting in what is known as a *false* alarm. A FDI scheme with high false alarm rates leads to a lack of confidence in the detection system and therefore the lower the false alarms the more robust the FDI scheme is to the unknown inputs. There have been several methods suggested in the literature which are specifically designed to improve the robustness of model-based FDI schemes such as; unknown input observers (which attempt to decouple the effects of unknown inputs on the residual) [105], eigenstructure assignment [65] and adaptive thresholds [106].

A strongly related performance criterion is the sensitivity of the FDI scheme to different fault types. The higher this sensitivity the lower the number of undetected faults and therefore the more reliable the fault detection system. Fault types which are generally difficult to detect include incipient faults (small and slow varying faults) which are typical of worn equipment [9]. Such faults can be difficult to detect as they initially cause minimal damage to the real system. To detect incipient faults, the FDI scheme must be adequately tuned. However this can also have the counter-effect of increasing the false alarm rates. For example lowering a residual threshold can increase the FDI scheme's sensitivity to incipient faults but it can also increase the false alarm rate. A trade-off is usually required between the desire to reduce the number of undetected faults and lowering the false alarm rate.

Finally, the ability to isolate (i.e. locate) the fault is important if appropriate maintenance action is to be designated. In other words fault accommodation is only possible if the fault is correctly isolated. Most of the work carried out has focused on FDI schemes. However the fault accommodation stage is equally important and often necessary. For example in aircraft a faulty sensor used in the control loops must be quickly replaced in order to avoid flight instability. In model-based FDI schemes the model estimate can replace the faulty sensor and overall such schemes are referred to as FDI and accommodation (FDIA) schemes.

2.3 Examples and Trends

For a complete introduction to FDI schemes, one must be aware of the research trends, i.e. the popularity of each method and its applications. In 1997, Isermann

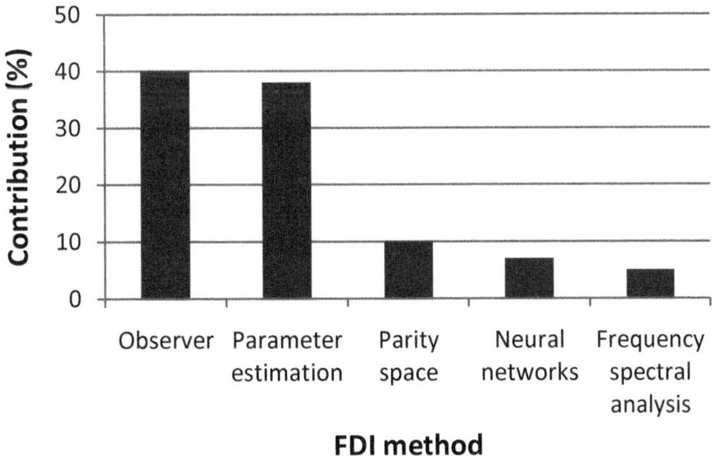

Fig. 2.4 Trends in FDI method [8]

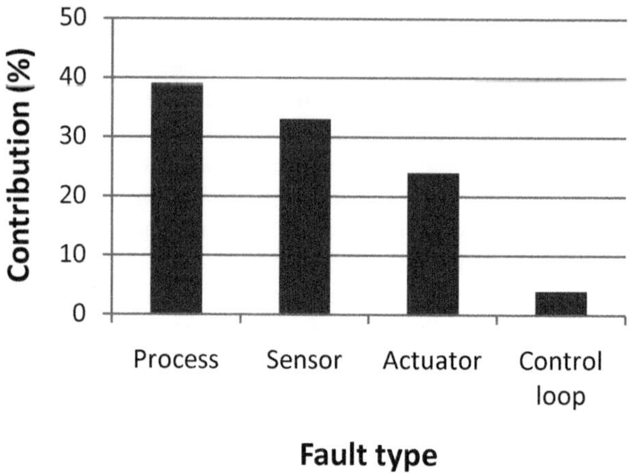

Fig. 2.5 Trends in fault type [8]

Table 2.1 Examples of model-based FDI applications published in the 21st century

Application	FDI method	Comments
Steam generator	Parity space	Sensor FDI of real steam generator data [107]
DC motor	Parity space	Sensor and actuator FDI applied to nonlinear simulation of DC motor [108]
Automotive (car)	Parity space	Sensor FDI using real driving data [109]
Aerospace (UAV)	Parity space	Actuator FDI demonstrated on nonlinear aircraft model [110]
Aerospace	Observer	Actuator FDI, using a multiple-model hypobook test approach, demonstrated on an aircraft model [111]
Thermoforming Process	Observer	Actuator FDI, using both dedicated and generalised Kalman filter schemes, applied to a nonlinear model [112]
Mechanical engineering (simply supported beam)	Observer (fault detection filter)	Structure fault detection using a fault detection filter, applied in simulation [113]
Satellite	Parameter estimation	Actuator FDI applied to a nonlinear model [114]
DC Motor	Parameter Estimation	Process FDI applied in both simulation and experimentally [115]
Automotive (car engine)	Neural networks	Sensor FDI using real sensor data from a Fiat engine [116]
Automotive (heavy-duty diesel engine)	Neural networks	Sensor FDI using real data obtained from a vehicle tested in urban and highway roads [117]
Aerospace	Neural networks	Sensor FDI using radial basis functions, applied to nonlinear aircraft model [118]

and Balle [8] published their famous paper which discussed such trends, some of which will be re-iterated here (Fig 2.4-2.5). From Fig 2.4, we note that observer-based methods have been more frequently applied. This could be as a result of the already well-established observer (and Kalman filter) theory. Almost 70% of FDI schemes use observer-based or parameter estimation methods while NN-based FDI schemes were rarely applied (Fig 2.4). However applications which make use of NNs have mainly targeted nonlinear applications. In fact the applications to linear processes were decreasing while nonlinear applications using NN-based methods were increasing in comparison to other FDI methods [8]. In Table 2.1 we show some of the FDI applications published in the 21st century.

Conclusions

This chapter is an introduction to FDI schemes with particular emphasis on; the terminology, model-based FDI methods, performance criteria and research trends. The efforts gone into the field of FDI is overwhelming. Despite this, there is yet to be a wide acceptance of FDI in industry which could be due to several reasons. Firstly it could be that traditional physical redundancy and limit-value checking approaches are simpler to implement. An additional reason is that current model-based methods are generally based on linear-time invariant models which can have very limited application in real systems. In contrast, NN-based FDI applications are gradually increasing due to their nonlinear structure and their ability to adapt to time-varying systems. In Chapters 5 and 6 a NN-based SFDIA scheme is designed and tested on an UAV application.

Chapter 3
Introduction to FADS Systems

Introduction

Traditionally, critical air data are measured using air data booms protruding from the aircraft local flow field; freestream static pressure (P_∞) and total pressure (P_0) are measured using a Pitot-static tube while angle of attack (α) and angle of sideslip (β) are measured using small vanes mounted on the air data boom. Using these four basic air data quantities ($P_\infty, P_0, \alpha, \beta$) as well as temperature (T_∞), most other air data of interest can be directly calculated such as; airspeed, altitude and rate of climb. In this book we are interested in measuring the critical air data; $P_\infty, P_0, \alpha, \beta$. Different designs and applications may exist, however the basic air data boom remains one of the most popular method for such air data measurements [35].

Despite their popularity, air data booms are known to have measurement disadvantages in addition to possible malfunctions: accuracy may be adversely affected by boom bending and vibration, probe size and geometry, and by the flow interference due to the probe itself. Furthermore, in military-related applications, external instrumentation is undesirable in stealth vehicles. As a result, in recent years more research has been carried out to find alternative solutions to air data booms. One example is optical air data system, which measures the atmosphere outside of an air vehicle and provides information regarding the environment ahead of the flight vehicle [119]. These systems are very accurate and more importantly are not affected by weather conditions external to the aircraft such as icing or plugging. However, with the primary goal of most air vehicle manufacturers being the reduction of costs, researchers have found the concept of air data measurements using a matrix of pressure orifices/ports to be a cheaper alternative to optical systems and booms.

The measurement of flush surface pressures to estimate air data parameters has been known for some time and is referred to as a Flush Air Data Sensing (FADS) system. The first FADS system was developed and tested on the NASA X-15 hypersonic aircraft [29, 30]. It consisted of a hemispherical nose (mounted with 4 pressure ports) which was steered into the wind vector to measure the air data. Results were promising, however the concept of the steered nose was considered too complex. Consequently, over the years the FADS system experienced many modifications and successful applications some of which will be discussed in section 3.2. Most aeronautical applications of the FADS system originate from the initial tests carried out by NASA in the early 1980s. Examples include [31-33]. Recently the FADS system was implemented on the NASA Dryden F-18 Systems

I. Samy and D.-W. Gu: Fault Detection and Flight Data Measurement, LNCIS 419, pp. 19–27.
springerlink.com

Research Aircraft [34]. This system uses 11 pressure ports in the radome of the aircraft and was tested at speeds up to Mach 1.6, α up to 80° and β= ±20°. Other applications of the FADS system include [35-37].

From the literature we find that few examples have been extended to mini (unmanned) air vehicles (MAVs). This motivated the investigation of such an application. MAVs are found within the spectrum of UAVs and are characterised by their small size and low weight. Several military missions have taken advantage of this feature to use them in applications which can only be attained at great risk. The FADS system is an invaluable alternative to air data booms especially for MAV applications. This is because current air data booms can be too heavy and expensive for use on a MAV. Additionally, due to the dangerous and secretive environments that they can be exposed to, external instrumentation is best avoided.

Most applications tend to mount the FADS system near the aircraft nosetip mainly for two reasons. Firstly the aerodynamic model relating the surface pressure and air data states is derived around a blunt body, and so is most valid at the nosecap which can be approximated as a sphere. Secondly the nosetip has been used traditionally as the air data measurement location for air data booms. Unfortunately many MAVs (as it is in our case) are driven by a nose-propeller which can obstruct the FADs system. Therefore as an alternative we consider placing the FADS system at the wing leading edge. In fact in [41] it was reported that the aerodynamic model developed for a FADS system mounted at the spherical nosecap, is equally applicable to the wing leading edge.

This chapter is organised as follows. In section 3.1 the popular air data boom is discussed and in section 3.2 the FADS system is introduced. Section 3.3 defines the FADS system model used to relate the aircraft surface pressure to the critical air data $P_\infty, P_0, \alpha, \beta$. Note that this chapter introduces the background and history of FADS systems while Chapter 7 presents an application to a MAV.

3.1 Air Data Boom

Air data are derived from the air surrounding the aircraft and include; indicated and true airspeed, pressure altitude, ambient air temperature, angles of attack and sideslip, Mach number and rate of climb [120]. Such air data can be directly calculated based on only five *critical* air data [35]:

- Freestream static pressure (P_∞): Pressure measured far ahead of the aircraft where the air is undisturbed from aircraft motion.
- Total pressure (P_0): Also referred to as the stagnation pressure. This is the pressure measured at a stagnation point in the airflow (i.e. where airflow is decelerated to approximately zero airspeed).
- Temperature (T_∞): Temperature of the atmosphere far ahead of the aircraft.
- Aerodynamic orientation: Includes aircraft angle of attack (α) and sideslip (β).

Fig 3.1 shows how these five basic air data quantities can be used to measure all other air data where [120]; indicated airspeed is the airspeed of the aircraft relative to a static atmosphere and assuming constant air density (i.e. incompressible flow), true airspeed is the indicated airspeed corrected for instrumentation errors and air density assumptions, Mach number is the ratio of true airspeed to the speed of sound in air, pressure altitude is the altitude of the aircraft based on the atmospheric pressure and the pre-defined International Standard Atmosphere look up table and rate of climb is the rate of change of the aircraft altitude.

In this book we are interested in only four of the five critical air data; $P_\infty, P_0, \alpha, \beta$. Traditionally these parameters are measured using what is known as an air data boom. An air data boom (Fig 3.3) can be divided into two components; a Pitot-static tube (Fig 3.2(a)) and a mechanical vane structure (Fig 3.2(b)).

The well known Pitot-static tube measures two pressures; P_0 and P_∞. To measure P_0, the front end of the Pitot-static tube is open to the air and directly faces the airflow (Fig 3.2(a)). This design acts as a direct blockage and therefore

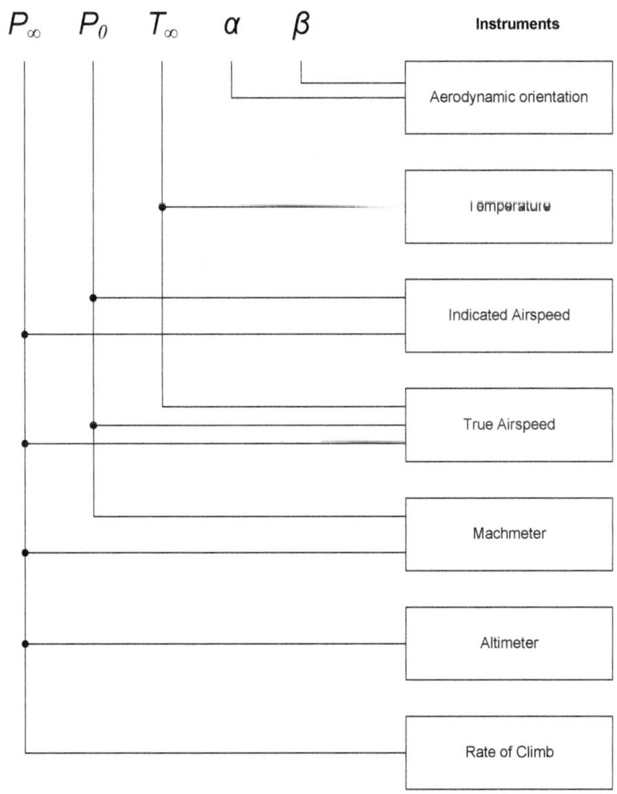

Fig. 3.1 The relation of the critical air data ($\boldsymbol{P_\infty, P_0, T_\infty, \alpha, \beta}$) to all other air data

decelerates the airflow to approximately zero airspeed. On the other hand the static port is used to measure P_∞ and is located at the tube surface, i.e. in parallel to the airflow (Fig 3.2(a)). Pitot-static tubes are the most popular method for calculating the aircraft speed [121]. The difference between P_0 and P_∞ can be used to calculate the indicated airspeed based on Bernoulli's equation which will be discussed in detail in Chapter 7. However note that P_0 and P_∞ can be individually used to calculate other data (e.g. pressure altitude can be calculated based on P_∞) and so it is important that P_0 and P_∞ are both measured and not just their difference.

The air data boom also consists of two mechanical vanes used to measure α and β. Fig 3.2(b) shows an example of such a vane. Vanes are aligned in the direction of zero α (or β) and if the aircraft aerodynamic orientation is perturbed the vanes pivot accordingly. The amount of rotation is then measured via a potentiometer.

The air data boom (Fig 3.3) must extend beyond the aircraft surface for two main reasons. Firstly, the Pitot-static tube needs to measure the freestream pressures, i.e. the pressure far ahead of the aircraft where the air is undisturbed by aircraft motion. Secondly the mechanical vanes must be located ahead of the aircraft since they should be the first part of the aircraft to be affected by the airflow. The latter condition can be justified by recalling that α and β are defined based on the angles the aircraft makes with the *freestream* airspeed vector. Therefore the mechanical vanes must be placed far ahead of the aircraft, i.e. at freestream conditions. Traditionally air data booms are located at the fuselage nose or wing tip.

Air data booms can come in different sizes and weights. For example, SpaceAge Control (popular air data boom manufacturer) produces air data booms (designed especially for UAVs) which can cost up to £2500 and weigh almost 170g (see 100400 air data boom for UAVs [123]). However the MAV used in this book (Chapter 7, section 7.1) is cheap to manufacture and weighs only 450g. Therefore traditional air data booms can often be impractical for MAVs and so FADS systems can be an invaluable alternative for measuring air data.

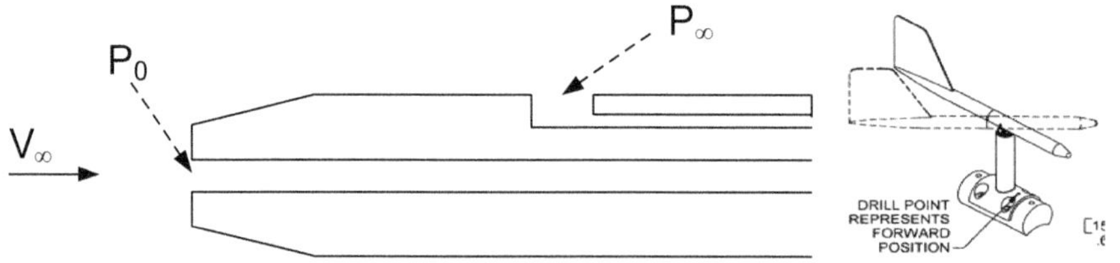

(a) (b)

Fig. 3.2 (a) Pitot-static tube, (b) Mechanical Vane (see 100386 mini vane [122])

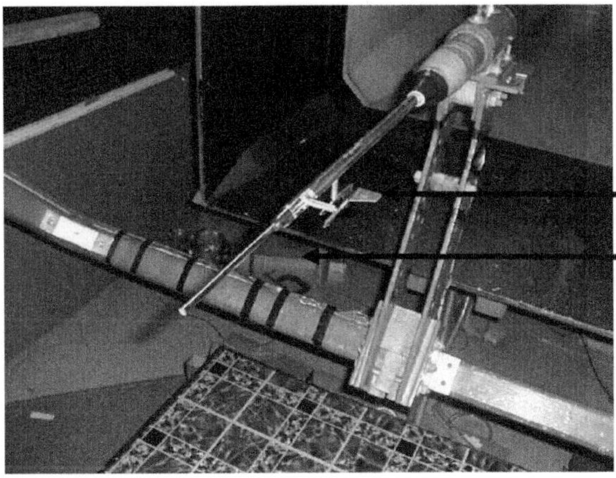

Fig. 3.3 Air data boom (Top view) [123]

3.2 Background and History

FADS systems convert aircraft surface pressure (measured from pressure orifices drilled into the aircraft surface) to air data such as $P_\infty, P_0, \alpha, \beta$. For reasons discussed earlier, this approach to air data measurement is preferred to traditional air data booms. Furthermore the FADS system can be more robust to faults in comparison to air data booms. For example if a FADS system consists of 100 pressure ports then a fault in one of the ports should not significantly degrade the air data estimation accuracy. On the other hand a blockage in the P_0 port (Fig 3.2(a)) of an air data boom completely eliminates the availability of a total pressure measurement. The fault tolerant property of FADS systems was noted and tested in [46, 36].

However the FADS system can also suffer several drawbacks. The fundamental purpose of an intrusive air data boom is to measure air data far ahead (i.e. freestream) of the aircraft local flow fields. On the other hand the pressure orifices of a FADS system are located on the surface of the aircraft and must therefore be carefully calibrated for any locally induced flow fields. Furthermore the pressure orifices are prone to blockage caused by e.g. icing effects and atmospheric debris. Therefore it is important that FADS systems are rigorously flight tested and compared to an air data boom prior to implementation.

The FADS system was developed for the NASA X-15 program [29]. In this case the pressure ports were located on a hemispherical nose which was actively steered through the air to measure air data. The mechanical design of this concept was deemed too complicated and was subsequently abandoned. In the 1980's a more advanced approach, the shuttle entry air data system, was developed for the space shuttle program at the NASA Langley Research Center [39]. The motivation behind this project was that a space shuttle experiences high temperatures (during the re-entry stage) which can destroy most air data booms. The non-intrusive FADS systems were therefore seen as an invaluable alternative. It consisted of 20 pressure ports *fixed* to the shuttle surface. This design avoided the complications of the steered nose concept developed for the X-15 program. Aeronautical applications of the FADS systems were also carried out at the NASA Dryden Flight Research Center [32, 33]. The FADS system tests proved that aircraft surface pressure can be related to the air data states. This relationship was empirically determined from wind tunnel tests and real flight tests and results were tabulated in terms of surface pressure ratios and the corresponding air data state [33]. However there was no attempt to define this relationship in terms of a mathematical model. It was not until the flight testing programs, carried out at NASA Dryden Flight Research Center, that such a model was developed [40, 41]. The FADS system was named the high angle attack FADS (HI-FADS) system as it was designed for an F-18 aircraft which was capable of reaching angle of attacks of more than 50 deg. Since then different research groups have proposed different design methods which differed in terms of e.g. pressure port location, number of pressure ports used and the FADS system model used [34-36, 38, 40-46].

3.3 FADS System Model

The relationship between the aircraft surface pressure and the air data $(P_\infty, P_0, \alpha, \beta)$ is known as the FADS system model or the air data model. The standard air data model used in the FADS system is defined in many parts of the literature [34, 38, 40-43]. It can be derived based on three airflow assumptions; irrotational (potential) flow, incompressible flow and airflow over a blunt body (e.g. sphere). Irrotational flow assumes that air simply translates over the body with no rotation, i.e. with zero angular velocity, while incompressible flow assumes that the air density is constant everywhere in the atmosphere. Blunt bodies have the property that most of the aerodynamic drag is caused by perpendicular forces instead of tangential forces [124]. For example a vertical plate can be categorised as a blunt body while a streamlined aerofoil is not a blunt body as most of the drag will be due to tangential ('tugging') forces. The blunt body assumed in the air data model is a sphere, as the pressure ports are generally located at the nosecap of the aircraft which has an almost spherical shape. Together these assumptions greatly simplify the air data model [43]. For a complete derivation of irrotational flow over a sphere the reader is referred to [124, 125].

The air data model can be defined as [43]:

$$p_i(\theta_i) = q_\infty[\cos^2\theta_i + \varepsilon\sin^2\theta_i] + P_\infty \tag{3.1}$$

where p_i is the surface pressure measured at port i, θ_i is the flow incidence angle between the surface normal at the i'th port and the airspeed vector, q_∞ is the freestream dynamic pressure (section 7.2.1) and ε is a calibration parameter which is empirically determined to account for the assumptions mentioned earlier (e.g. incompressible flow). The angle θ_i can be defined in terms of the local angle of attack and sideslip $(\alpha_{eff}, \beta_{eff})$ as follows [43]:

$$\cos(\theta_i) = \cos(\alpha_{eff})\cos(\beta_{eff})\cos(\lambda_i) + \sin(\beta_{eff})\sin(\phi_i)\sin(\lambda_i) + \sin(\alpha_{eff})\cos(\beta_{eff})\cos(\phi_i)\sin(\lambda_i) \quad (3.2)$$

where λ_i is the angle the normal to the surface at port i makes with the longitudinal axis of the nosecap and the clock angle (ϕ_i)is the clockwise angle looking aft around the axis of symmetry starting at the bottom of the nosecap (Fig 3.4). Combining (3.1) and (3.2) we can define the overall air data model relating a vector of surface pressure measurements $\mathbf{p_i}$, to the air data states as:

$$\mathbf{p_i}(\phi_i, \lambda_i) = \boldsymbol{f}_i(P_\infty, P_0, \alpha_{eff}, \beta_{eff}, \varepsilon, \phi_i, \lambda_i) \quad (3.3)$$

where i=1,2,....N and N is the total number of pressure ports in the FADS system. The air data model (3.3) is also applicable if the FADS system is located at the wing leading edge [38].

Equation (3.3) is quite complex and can be difficult to solve. Furthermore, α_{eff} and β_{eff} are local flow angles influenced by the aircraft-induced wash [35]. Therefore they must be calibrated so that the true freestream aerodynamic orientation (α, β) is calculated. To further complicate (3.3), the calibration parameter ε is itself dependant on the air data (e.g. ε would change with airspeed). There are two popular methods, cited in the literature, to solve (3.3); nonlinear regression methods and the triples algorithm. The nonlinear regression approach substitutes the measured $\mathbf{p_i}$ into (3.3) and air data states are then estimated using a nonlinear least squares regression algorithm. Applications of such a method can be found in [34, 40, 41] This method has known to suffer from stability problems (i.e. solutions can often diverge) but more importantly can be computationally demanding due to its iterative approach. In fact the same authors who proposed this method for solving (3.3), acknowledged that it was too risky to be used in real time and was subsequently abandoned [43]. As an alternative however they proposed the triples algorithm [43]. The triples algorithm is a clever way to solving (3.3). It strategically selects 3 pressure ports to decouple $\alpha_{eff}, \beta_{eff}$ from $P_\infty, P_0, \varepsilon$. Therefore it is important that λ_i and ϕ_i are carefully selected. In other words the pressure ports are not randomly distributed over the aircraft surface, but are instead placed at specific angles (λ_i and ϕ_i) so that some terms in (3.1) and (3.2) are cancelled out. So for example if we place a pressure port at $\lambda_i = 0\ (or \pm 180\ \text{deg})$ then we can see that the term in (3.2) is greatly simplified. The triples algorithm can be difficult to implement in MAV applications as the exact pressure port locations cannot be accurately defined in such small air vehicles. An application of the triples algorithm can be found in [43].

The air data model defined in (3.3) is based on several assumptions (such as spherical shapes) and is therefore susceptible to modelling errors. Simpler approaches to modelling (3.3) are via lookup tables, where calibration data (from either wind tunnels tests or real flight) is used to build the lookup table. This is the simplest of methods but it also suffers from high memory usage and slow execution times.

As an alternative, we investigate the feasibility of using NNs to model the air data system model. NNs provide a means of modelling linear or nonlinear systems without the need of detailed system knowledge. They primarily rely on sufficient training data from which they can develop their structure. This makes them an attractive modelling solution to applications where the theory is poor but the relevant data is plentiful. An advantage of NNs is that it is possible to implement them in just a few lines of code which is suitable in MAV applications where computational power may be limited. A drawback however is that its structure is purely based on available system input/output data. Therefore the training algorithm and its implementation is a crucial step in NN design. The two most popular NN architectures are the multilayer perceptron (MLP) and the radial basis function (RBF) NN. MLP NNs trained with error backpropagation algorithms have been successfully implemented in a FADS system by several authors [35, 36, 44, 45]. Instead, we consider a RBF NN trained with the extended minimum resource allocating network (EMRAN) algorithm (Chapter 4).

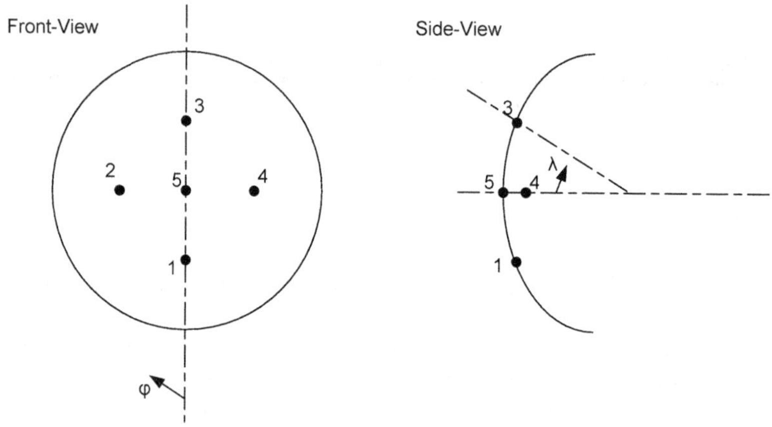

Fig. 3.4 Definition of cone (**λ**) and clock angle (**φ**) for a spherical nosecap with 5 pressure ports

Conclusions

In this chapter we have discussed the background of FADS systems. Flight data monitoring systems are an important part of aircraft and can be used for many purposes such as air traffic control, flight performance analysis, health monitoring, post air-crash analysis etc. In this book we are interested in a section of the flight data known as the critical air data ($P_\infty, P_0, \alpha, \beta$). Traditionally such air data are measured using an air data boom protruding from the aircraft. However new research trends have shown that air data computed from a matrix of pressure orifices distributed on the surface of an aircraft can be a cheaper alternative to booms. FADs systems convert aircraft surface pressure to the critical air data states. Moreover they are an invaluable solution to air data measurement in MAVs due to the cost and weight implications associated with traditional air data booms. It was also noted that the critical air data estimated from the FADS system along with temperature (T_∞) are sufficient to calculating any other air data quantity of interest such as Mach number, true airspeed, pressure altitude and rate of climb. The standard FADS system model was also defined and popular methods for solving such a model were discussed. However it was noted that such models are derived based on several assumptions (e.g. spherical shapes, incompressible flow) and can be quite difficult to solve. Alternatively the benefits of using NN models to relate the surface pressure to the critical air data ($P_\infty, P_0, \alpha, \beta$) have been suggested. A NN-based FADS system will be designed and tested on a MAV in Chapter 7.

Chapter 4
Neural Networks

Introduction

In 1943, McCulloch and Pitts first proposed the feedforward perceptron and since then neural networks have found their way into a variety of applications [103, 104]. Artificial neural networks (NNs) consist of a large number of simple processing elements called neurons which are interconnected together via channels called connections. This structure is in fact inspired by the structure of the brain's biological nervous system [103].

One of the benefits of NNs is their highly interconnected structure which can make them fault tolerant, i.e. performance is not significantly degraded if one of its links or neurons is faulty. For example if thousands of inputs are connected to only one neuron then the output of this neuron is likely to be robust to faults in one or more of its inputs. However the main property of NNs which has often made them superior to traditional modelling methods is their online adaptive capabilities. Using an appropriate training algorithm a NN can update its structure, in real time, to better suit the input/output data. In particular, a NN trained to operate in a specific environment can be easily re-trained to deal with minor changes in the operating conditions. This property is especially useful when modelling time-varying systems. For a general introduction to NNs and their developments, since McCulloch and Pitts first proposed the feedforward perceptron in 1943, the reader is referred to [97, 103, 104, 126].

Due to their highly interconnected structure it can be difficult to visualise and predict the performance of a NN [97]. It is therefore important that the NN model chosen is rigorously tested before implementation. Due to their nonparametric modelling approach, they are best suited in applications were the theory is poor but the training data is plentiful.

NNs are considered to be an interdisciplinary field due to their application in a wide variety of fields. For example Faro *et* al. make use of NNs for traffic monitoring purposes [127]. Using inputs from humans and webcams the NN computes important traffic parameters such as traffic flow, density, and car speeds etc. [127]. [128] use NNs in the field of finance. Using past gold prices and past Dow Jones indices, the NNs are designed for one-step predictions of the price of gold. Each NN is a multilayer perceptron (MLP) and real gold price data is used to train and test the NNs. In a different application, Wang *et* al. make use of NNs in space research [129]. A MLP NN (3-3-1 structure) trained with the conventional backpropagation algorithm is used to model the solar activity of the sun. The model receives information regarding the solar magnetic field properties of the

I. Samy and D.-W. Gu: Fault Detection and Flight Data Measurement, LNCIS 419, pp. 29–35.
springerlink.com

sun and estimates its solar productivity. The concept is tested using real satellite data. Further industrial applications of NNs can also be found here [130].

In our case the NNs are used as nonlinear function approximators in the SFDIA schemes and the FADS system. In the SFDIA scheme (Chapters 5-6) the NNs model the relationship between a group of sensors while in the FADS system (Chapter 7) the NN is used to model the functional relationship between wing surface pressure and air data (e.g. airspeed, angle of attack, sideslip). There have been several publications which have made use of NNs in SFDIA schemes and FADS systems. Examples of these include the work published by the group of Napolitano which have demonstrated the use of NN-based SFDIA schemes in large manned aircrafts [118, 24-28]. MLP NNs have also been successfully implemented in FADS systems [35, 36, 44, 45]. However few applications have been extended to UAVs.

4.1 NN Structure and Training

4.1.1 RBF NN

The two most popular NN architectures are the MLP NN and the RBF NN [97]. Over the past years there has been a particular interest in RBF NNs due to their good generalisation performance [131, 132]. Good generalisation performance is important in SFDIA schemes and FADS systems as NN training is eventually switched off (i.e. the NN structure is frozen) and therefore the accuracy of the future NN estimates greatly depends on the NN's ability to generalise to 'new' data.

A typical 2-layered RBF NN consisting of the inputs, a hidden layer and an output layer is shown in Fig 4.1. The inputs [*x1, x2…,xn*] are not counted as a layer as no computations are performed. Also note that weights are not found between the inputs and the hidden layer. The hidden layer consists of the nonlinear RBFs which form a localised response to the input vector [*x1 x2…xn*], based on the distance between the input vector and the RBF centres. The smaller this distance the larger the output of the hidden neuron (i.e. the higher the activation) and in principle each hidden neuron can be tuned to be sensitive to only one specific input pattern. Theoretical analysis and practical results have shown that the choice of nonlinearity in the hidden layer is generally not crucial to the performance of the RBF NN [133]. Gaussian functions are typically used. The output layer then performs a simple linear combination of the hidden layer outputs (Fig 4.1).

A single output RBF NN with N hidden neurons can be expressed as follows:

$$\hat{y} = \lambda_0 + \sum_{n=1}^{N} \lambda_n \exp\left[\frac{-||\mathbf{x} - \boldsymbol{\mu}_n||^2}{\sigma_n^2}\right] \tag{4.1}$$

where $\hat{y}$ is the NN estimate of target y, $\mathbf{x}$ is the input vector, λ are the individual weights found between the hidden and output layer, and $\boldsymbol{\mu}$ and σ are the centres and widths of the Gaussian functions (hidden neurons) respectively. ||.|| is the Euclidean norm.

In the conventional implementation of the RBF NN, N in (4.1) is fixed and assumed to be known *a priori*. The training algorithm is then used to update only the weights of the RBF NN, i.e. the centres and widths of the Gaussian functions are fixed. This approach is appealing as well-established linear optimisation methods such as the least mean square (LMS) algorithm could be used to update the linearly connected weights.

However fixing the centres and widths of the hidden neurons has shown to result in large NN structures (as in principle we would require one hidden neuron for each input pattern). This dimensionality problem can significantly increase the NN processing time [133]. Moreover, heuristically selecting a suitable number of hidden neurons as well as their respective centres and widths can be a challenging task. To overcome the drawbacks of conventional RBF NNs, a more advanced network was developed by Li et *al.* in [132] known as the EMRAN RBF NN which is in fact an extension to the RAN RBF NN originally proposed by Platt in 1991 [134].

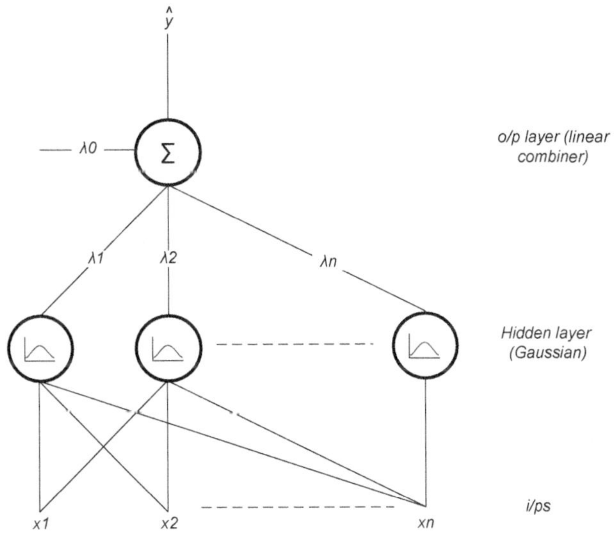

Fig. 4.1 Fully connected RBF NN. x, λ, $\hat{\mathbf{y}}$ are inputs, weights and output respectively

4.1.2 EMRAN RBF NN

The RAN algorithm proceeds as follows (only considering a single output RBF NN as in (4.1) and Fig 4.1). Initially the RBF NN starts with no hidden neurons and $\lambda_0 = y_0$, where y_0 is the target output at sample $k = 0$. It then adds hidden neurons **if all three** of the following criteria are met:

$$e_k = y_k - \hat{y}_k \qquad > E1 \tag{4.2}$$

$$e_{RMSk} = \sqrt{\sum_{j=k-(M-1)}^{k} \frac{e_j^2}{M}} > E2 \tag{4.3}$$

$$d_k = ||\mathbf{x_k} - \boldsymbol{\mu}_{\mathbf{rk}}|| > E3 \tag{4.4}$$

where $\boldsymbol{\mu}_{\mathbf{rk}}$ is the centre of the hidden neuron closest to the input vector $\mathbf{x_k}$. *E1*, *E2* and *E3* are fixed thresholds and $E3 = \max\{\varepsilon_{max}.\gamma_{dy}^k, \varepsilon_{min}\}$ where $\gamma_{dy}$is a positive decay constant ($0 < \gamma_{dy} < 1$). *E1* and *E2* ensure that the estimation error and the root mean square (RMS) estimation error for the past *M* samples are below a pre-defined threshold, i.e. they check if the NN estimates are sufficiently accurate. *E3* checks if the minimum distance between the current input vector and the centers of the hidden neurons is significantly small (i.e. so that there is at least one hidden neuron which is sensitive to the current input vector pattern). If all three criteria in (4.2)-(4.4) are satisfied then a new hidden neuron with the following properties is added:

$$\lambda = e_k \tag{4.5}$$

$$\mu = \mathbf{x_k} \tag{4.6}$$

$$\sigma = k_{op}||\mathbf{x_k} - \boldsymbol{\mu}_{\mathbf{rk}}|| \tag{4.7}$$

Specifying the weight and centre as in (4.5)-(4.6) allows us to remove the error e_k experienced at sample instant *k*, i.e. e_k (not e_{k+1} as this would then depend on y_{k+1}) would equal to zero if we would have included a hidden neuron as in (4.5)-(4.7) at sample instant *k*. The overlap between the new hidden neuron and the hidden neuron with centre $\boldsymbol{\mu}_{\mathbf{rk}}$ is based on (4.7) where k_{op} is the overlap factor.

If **less than three** of the criteria in (4.2)-(4.4) are met then a new hidden neuron is *not* added (i.e. it is assumed that the NN performance is not significantly poor) and instead a chosen training algorithm (section 4.1.3) updates all existing free parameters; i.e. centres (μ), widths (σ) and weights (λ). The resulting NN is referred to as a RAN RBF NN and was originally developed to overcome the dimensionality problems associated with the conventional RBF NN [134].

This can be extended to a MRAN RBF NN by pruning (removing) the hidden neurons which contribute the least to the network output and substituting them with more appropriate hidden neurons. However this is only implemented once the maximum number of hidden neurons allowed (N_{max}, defined by the user) is reached. The process of pruning is as follows. If the three criteria (4.2)-(4.4) are met and $N = N_{max}$ then the MRAN algorithm adds a new hidden neuron **in place of** the hidden neuron with the lowest activation (i.e. lowest output).

A drawback of the MRAN RBF NN is that all free parameters are updated during training which can be time consuming especially if there are many hidden neurons, i.e. N_{max} is large.

Consequently [132] proposed the EMRAN RBF NN which updates the free parameters of only one 'winner' neuron in an attempt to speed up the training process while maintaining the same approximation characteristics of the MRAN

RBF NN. The neuron with the centre closest to the data input vector $\mathbf{x_k}$ is chosen as the 'winner' neuron as it will probably have the highest activation, i.e. will contribute the most to the NN output.

4.1.3 NN Training Algorithm

The EMRAN algorithm is a set of conditions which decide how the RBF NN structure should be adapted to better suit the training data. However we are yet to define a training algorithm to update (only when directed by the EMRAN algorithm) the NN free parameters. Good characteristics of the chosen training algorithm include; faster execution times, lower estimation errors, and a compact structure i.e. if the training algorithm is poor then more hidden neurons may be needed. Most NN training algorithms are based on gradient descent optimisation methods where the NN free parameters are updated in a way which minimises the output estimation error. For example the standard LMS algorithm is typically used in the conventional RBF NN (section 4.1.1) to update the linearly connected weights. This technique would benefit from linear optimisation but as we have mentioned earlier, the NN can sometimes suffer from the curse of dimensionality. Other popular examples are the error back-propagation algorithm (BPA) and its extension, the extended BPA (EBPA) which are both used to train the famous MLP NN. However in this case, as the MLP NN has nonlinear neurons in its output layer and hidden layer, the process of training is a nonlinear optimisation problem and as such there are risks of being trapped in local minima. There have also been examples where an extended Kalman filter (EKF) is used to update the NN free parameters [135].

The gradient descent training algorithm often used to train the EMRAN RBF NN is as follows [118], Fig 4.2:

$$\boldsymbol{\theta}_{\mathbf{k+1}} = \boldsymbol{\theta}_{\mathbf{k}} + \Delta\boldsymbol{\theta}_{\mathbf{k}} \tag{4.8}$$

$$\Delta\boldsymbol{\theta}_{\mathbf{k}} = -\delta \left.\frac{\partial E}{\partial \boldsymbol{\theta}}\right|_k \tag{4.9}$$

where $\boldsymbol{\theta}$ is the vector of free parameters (made up of centres, widths and weights), δ is the learning rate and E is the instantaneous squared error (cost function):

$$E = \frac{1}{2}e^2 \tag{4.10}$$

where e is the NN estimation error as in (4.2). During each iteration the NN free parameters are adjusted in a direction opposite to the gradient in (4.9), i.e. towards the minimum squared error. The hidden layer of the EMRAN RBF NN consists of nonlinear neurons (Gaussian functions) and therefore the gradient descent training algorithm in (4.8) is a nonlinear optimisation problem i.e. we run the risk of being trapped in local minima. As such the NNs used in the SFDIA scheme and FADS system must be rigorously tested and the parameters $[E1, E2, \varepsilon_{max}, \varepsilon_{min}, \gamma_{dy}, M, k_{op}]$ (4.2)-(4.7), must be carefully tuned until we achieve acceptable performance (e.g. low estimation errors and fast execution times). This is also the case when choosing the NN learning rate as a high learning rate guarantees good estimations but it also degrades the global approximation capability of the NN.

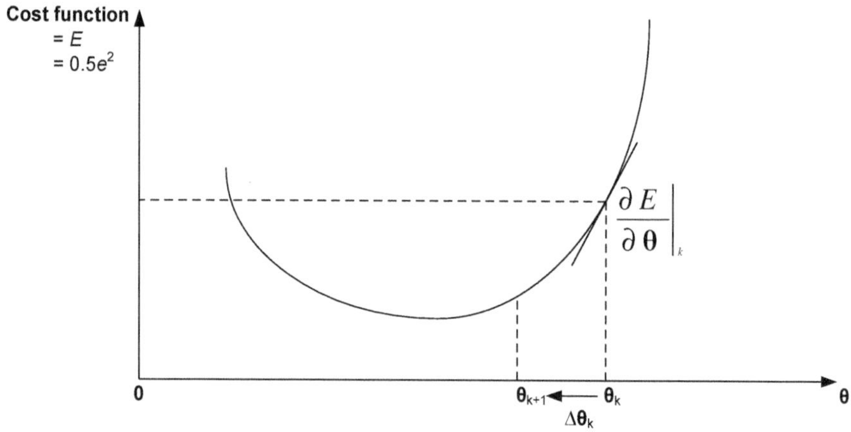

Fig. 4.2 Error surface showing the method used to train the EMRAN RBF NN

4.2 Application to the SFDIA Scheme and FADS System

The NNs used in the SFDIA scheme and FADS system are trained based on the estimation error e (4.2), i.e. the free parameters are updated in such a way which reduces the estimation error (see training algorithm, section 4.1.3). Therefore the desired (i.e. *target*) NN outputs must be defined *a priori* so that the estimation error e can be calculated at each time step. This method of training is more formally known as supervised training. Supervised training can be either implemented *online* or *offline*. In offline training the NN is trained and once satisfactory performance (e.g. by judging the accuracy and generalisation capability of the NN) the NN structure is *frozen* and used on board the aircraft with no further training. On the other hand online training continuously updates the NN structure in real time and is generally preferable in time-varying systems. However online training can also be computationally slow. Moreover it requires additional instrumentation on board the aircraft to measure the target outputs, which can be inadequate in applications where instrumentation costs are to be reduced.

The methods used to train the NNs in the SFDIA scheme and FADS system are as follows:

- **SFDIA scheme:** Both offline and online training. Offline training is carried out to *initialise* the NN structure. Once pre-defined stopping criteria (to be defined later) are satisfied, offline training is stopped and the NN is used in the SFDIA scheme. During the SFDIA application, the NN is continuously trained (i.e. online training) using the available sensor measurements as the target outputs.
- **FADS system:** Only offline training. In our case the purpose of the FADS system is to provide an alternative way to measuring the air data in a MAV, i.e. to avoid using an air data boom. To be able to train the NN online (i.e.

on-board the aircraft) we would require an extra air data sensor (such as an air data boom) to provide the target outputs. This therefore contradicts the purpose of the FADS system. As such, the NN is trained offline and once pre-defined stopping criteria are satisfied, the NN structure is frozen and used on-board the MAV with no further training.

A common problem with NNs is that of over-fitting (overtraining) the training data. In this case the NN estimations are accurate during the training process but are poor when exposed to 'new' data, i.e. the NN generalisation capabilities are poor. To overcome this problem, it is common practice to query the NN using a 'testing' data set as it is trained. NN training can then be stopped once the test set error starts to increase (as this indicates that the NN has been overtrained). In our case the offline training stopping criteria chosen are based on checking NN convergence as well as avoiding the over-fitting phenomenon. Two data sets are used to do so:

- Train data set: This is used to train the NN with learning switched *on*.
- Test data set: This is used to query the NN with learning switched *off*.

NN offline training is then stopped based on two criteria:

- Criterion 1: If the RMS estimation error of the test data set increases for more than 100 consecutive epochs.
- Criterion 2: If the rate of change of RMS per epoch (ΔRMS) is less than 0.1% for more than 100 consecutive epochs for both the test and train data sets [97].

One epoch represents one pass through the whole data set, Criterion 1 is the over fitting criterion and Criterion 2 checks if the NN has converged. Note that in the SFDIA scheme the offline training stage initialises the NN structure while in the FADS system, the offline training stage builds the final NN structure to be used on-board the aircraft.

Conclusions

In this chapter, we have presented the NN structure to be used in the SFDIA scheme and the FADS system. The EMRAN RBF NN is chosen due to its good generalisation capabilities, compact structure and fast execution times as reported in several parts of the literature (see e.g. [24, 118, 132]). One of the reasons for choosing a NN model is their online adaptive capabilities in comparison to traditional approaches which rely on a fixed mathematical model, such as Kalman filters. The online training capabilities of the NN will be investigated in the following chapters, and in Chapter 5, the performance of the NN in comparison to the Kalman filter is also considered. The method used to train the NN has also been discussed in this chapter and it was noted that online training in the FADS system is not practically possible due to the requirement of additional instrumentation such as an air data boom. Finally, the NN training stopping criteria were defined based on the need to check for NN convergence and to avoid over-fitting the NN structure.

Chapter 5
SFDA-Single Sensor Faults

Introduction

In this chapter, a NN-based and EKF-based sensor fault detection and accommodation (SFDA) scheme are proposed and compared under different levels of unknown inputs and fault types. The schemes are tested on a nonlinear UAV model. The EKF is chosen as a representative of nonlinear (fixed) model-based SFDA schemes which rely on a mathematical description of the real system. On the other hand, the NN is chosen due to its adaptive structure and online training capabilities. To test their robustness to unknown inputs, different levels of system and measurement noise are considered in the UAV model. Parameter uncertainties are also included in the EKF equations to investigate the performance of such methods to modelling errors.

The sensitivities of the SFDA schemes to the type of fault are also investigated. Fault types include; constant bias, step type, hard and soft additive faults. In this chapter, it is assumed that the sensor can fail only once and the faults are only present in the pitch gyro (Chapter 6 considers multiple sensor fault scenarios). Furthermore in an attempt to reduce the false alarm rates and number of undetected faults a novel residual generator is proposed which will be referred to as residual generation, padding and evaluation (RGPE). RGPE is compared to a standard residual generator which will be referred to as residual generation and evaluation (RGE).

The NN design chosen is the EMRAN RBF NN discussed in Chapter 4 and tests are implemented in a Matlab/Simulink environment. To the author's knowledge, the only work similar to the one presented here, is by Napolitano *et.* al [136]. In [136] a MLP NN trained via the EBPA is compared to a KF at linear conditions. However the same author suggested in [24] that the EMRAN RBF NN generally outperforms the MLP NN in terms of estimation accuracy and generalisation capabilities [24].

This chapter is organised as follows. In section 5.1 we introduce the notations and terminologies used throughout the chapter as well as the general outline of model-based SFDA schemes. In sections 5.2 and 5.3 we discuss the real UAV and its nonlinear model. In section 5.4 the EKF is discussed and in section 5.5 we introduce the RGE and RGPE methods. The input/output structure of the NN and EKF used in the SFDA schemes is presented in section 5.6. In section 5.7 we define the sensor fault types which are considered in the SFDA tests. In section 5.8 the SFDA test conditions and procedures are outlined in detail followed by the

I. Samy and D.-W. Gu: Fault Detection and Flight Data Measurement, LNCIS 419, pp. 37–82.
springerlink.com

Results, Discussion and Conclusion sections. Note that all results (tables, graphs etc.) are grouped together at the end of the chapter to facilitate their analysis.

5.1 General SFDA Outline and Terminologies

The following notations are used in this chapter. Sensor 'x' is referred to as Sensor-x. For example the pitch gyro is referred to as Sensor-q. The sensor measurements (i.e. UAV model outputs) are denoted by a subscript '*real*'. For example the Sensor-q measurements are referred to as q_{real}. However in some cases we may drop the subscript '*real*' for simplicity purposes. The estimate of these measurements are denoted by '$\widehat{\ }$' and a subscript which depends on the method of estimation. For example if a NN or EKF is used, then the estimate of q_{real} is referred to as $\hat{q}_{NN}$and $\hat{q}_{EKF}$ respectively. Finally the ideal sensor measurement (i.e. the measurement if no faults are present) is denoted by a subscript '*ideal*'.

In general SFDA of single sensor faults can be divided into two stages. Consider Fig 5.1 where only Sensor-y1 can be faulty. Stage 1 involves system modelling where $y1_{real}$ is estimated to give $\hat{y}1$ using the available *m-1* sensor measurements ($y2_{real}$, $y3_{real}$,…,ym_{real}). Stage 2 involves residual processing which consists of a residual generator and simple threshold logic. Fault detection and fault accommodation are implemented as follows:

- **Fault detection:** A fault alarm is triggered if the residual exceeds its threshold.
- **Fault accommodation:** If a fault is detected, the faulty sensor is replaced with the model estimate. Therefore in Fig 5.1 the switch would shift so that the output is now $\hat{y}1$, where ideally $\hat{y}1 = y1_{ideal}$

In our application, the sensor can only fail once, and this failure is permanent. Therefore in Fig 5.1 the SFDA scheme is terminated once the fault is detected and accommodated. For analysis purposes we can then compare $\hat{y}1$ and $y1_{ideal}$ to judge the NN fault accommodation performance. In our tests, 4 SFDA schemes are compared:

- **NN-RGE**
- **NN-RGPE**
- **EKF-RGE**
- **EKF-RGPE**

For example, the notation NN-RGE indicates that a NN is used for system modelling, and the RGE approach is used for residual processing. Note that in NN-based SFDA schemes, NN training is switched off once a fault is detected to avoid learning faulty measurements.

To summarise the notations used in this chapter consider the following statement:

"In this chapter four SFDA schemes are implemented, NN-RGE, NN-RGPE, EKF-RGE, EKF-RGPE. Faults in Sensor-q are only considered and the residual compares q_{real} *and* $\hat{q}_{NN}$ *(or* $\hat{q}_{EKF}$*). In the event of fault detection* $\hat{q}_{NN}$ *(or* $\hat{q}_{EKF}$*) replaces* q_{real}*, for fault accommodation purposes, and in the ideal scenario* $\hat{q}_{NN} = q_{ideal}$*."*

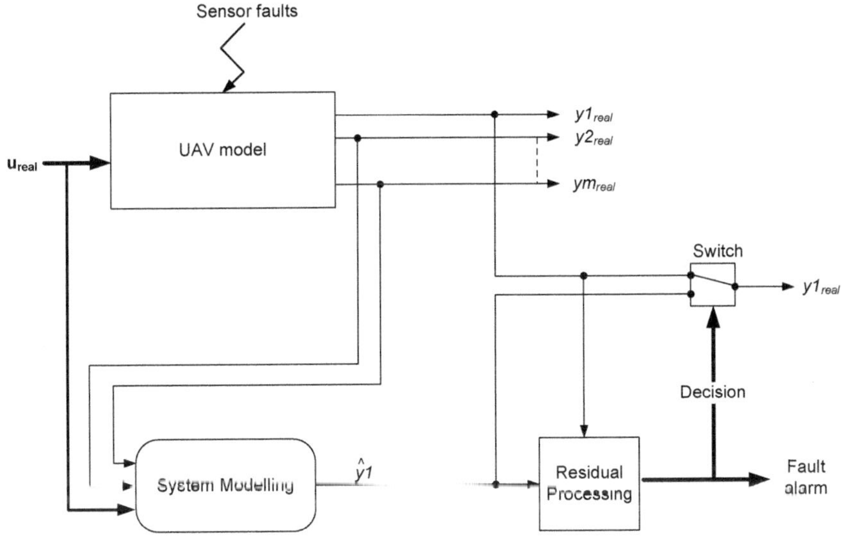

Fig. 5.1 General SFDA outline for a fault in Sensor-y1

5.2 UAV Used in the SFDA Schemes

The UAV is based on the Eclipse class vehicle [3] and is powered by a small gas turbine engine. It has three wing trailing edge devices (A, C, D) either side of the centre line (Fig 5.2):

- A: aileron type device (for roll control)
- B: fixed surface
- C: flap type device
- D: elevator type device (for pitch control)

(The fixed surface B is included for other parts of the project which are not considered here). A typical rudder type device is also located at the trailing edge of the fin for yaw control. The basic wing configurations and other properties are defined in Table 5.1.

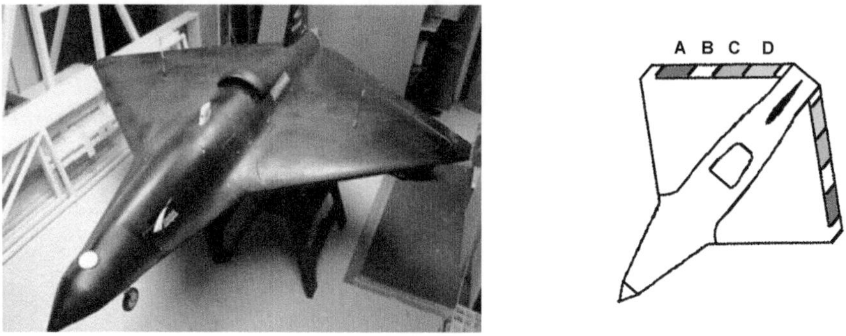

Fig. 5.2 The UAV and location of the control surfaces

Table 5.1 UAV configurations

Component	Symbol	Value
Wing area	S	2.365 m^2
Wing span	b	2.2 m
Mean aerodynamic chord	$\bar{\bar{c}}$	1.34 m
Aspect ratio	AR	2.047
Mass	m	39.7 kg
Centre of gravity location (with respect to nose)	x_{cg}	1.217 m
Roll inertia	I_x	1.5 kgm^2
Pitch inertia	I_y	10.43 kgm^2
Yaw inertia	I_z	11.41 kgm^2
Product inertia	I_{xz}	-0.14 kgm^2

5.3 UAV Model

A nonlinear decoupled, 6DoF model (open-loop, i.e. no stability augmentation) of the UAV was designed at Cranfield University and implemented in a Simulink environment [4]. Aircraft motion is defined by a body axis system (origin at aircraft centre of gravity) with respect to a fixed (inertial) earth axes (Fig 5.3). The UAV is assumed to be symmetric and rigid, and flight dynamics are described by the standard twelve first order differential equations; force, moment, kinematics and navigation equations. The twelve state variables include the components of velocity (u, v, w), angular rates (p, q, r), attitude (Φ, θ, ψ) and position (P_n, P_e, h) relative to the earth axes.

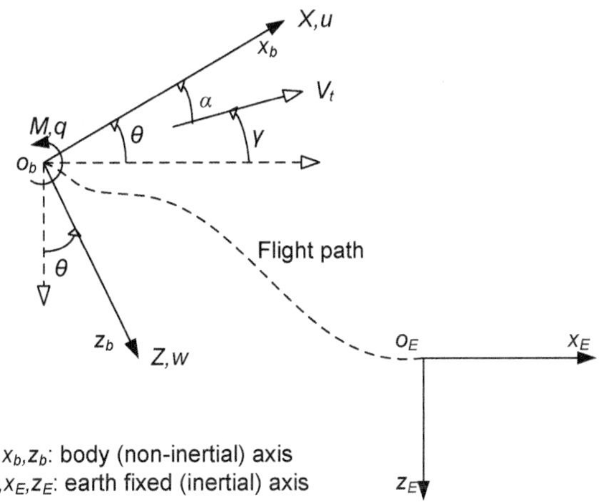

Fig. 5.3 UAV orientation and relevant motion variables (shown in the x-z plane)

5.3.1 Longitudinal Equations of Motion

UAV longitudinal motion is only considered in the SFDA tests; zero roll, yaw and sideslip angles (Φ, ψ, $\beta = 0$). The remaining lateral motion variables are also zero ($v = p = r = 0$) and aircraft motion is fully described by the axial force X, normal force Z and pitching moment M (Fig 5.3). The decoupled longitudinal equations of motion can be described as (Note: the lateral equations of motion can be found in [4]):

$$\begin{bmatrix} \dot{u} \\ \dot{w} \\ \dot{q} \\ \dot{\theta} \\ \dot{P}_n \\ \dot{h} \end{bmatrix} = \begin{bmatrix} (X_{aero} + X_{thrust} + X_{gravity})/m - qw \\ (Z_{aero} + Z_{thrust} + Z_{gravity})/m + qu \\ (M_{aero} + M_{thrust} + M_{gravity})/I_y \\ q \\ u\cos\theta + w\sin\theta \\ u\sin\theta - w\cos\theta \end{bmatrix} \tag{5.1}$$

X, Z and M are assumed to be due to aerodynamic, power and gravitational effects and are defined as follows:

$$\begin{bmatrix} X_{aero} \\ Z_{aero} \\ M_{aero} \end{bmatrix} = \begin{bmatrix} 0.5\rho V_t^2 S(C_D \cos\alpha - C_L \sin\alpha) \\ 0.5\rho V_t^2 S(C_D \sin\alpha + C_L \cos\alpha) \\ 0.5\rho V_t^2 S\bar{\bar{c}} C_M \end{bmatrix} \tag{5.2}$$

$$\begin{bmatrix} X_{thrust} \\ Z_{thrust} \\ M_{thrust} \end{bmatrix} = \begin{bmatrix} \tau T(V_t, h) \\ 0 \\ -X_{thrust} z_t \end{bmatrix} \tag{5.3}$$

$$\begin{bmatrix} X_{gravity} \\ Z_{gravity} \\ M_{gravity} \end{bmatrix} = \begin{bmatrix} -mg\sin\theta \\ mg\cos\theta \\ 0 \end{bmatrix} \tag{5.4}$$

where ρ is the air density, V_t is the airspeed, α is the angle of attack, τ is the engine throttle setting, z_t is the z-axis coordinate of the engine thrust line (which is parallel to the ox_b axis, Fig 5.3) relative to aircraft body axis. The lift, drag and pitching moment coefficients in (5.2) are functions of the following variables:

$$\begin{bmatrix} C_L \\ C_D \\ C_M \end{bmatrix} = \begin{bmatrix} C_L(\alpha,\eta) \\ C_D(\alpha,\eta) \\ C_M(\alpha,\eta,q,\dot{w}) \end{bmatrix} \tag{5.5}$$

The engine output thrust is assumed to act only along the axial body axis of the aircraft and is described by a simple relationship between throttle setting (τ) and available thrust, T_X:

$$T_X = \tau \mathrm{T}(V_t, h) \tag{5.6}$$

where the available thrust is described by the lookup table $\mathrm{T}(V_t, h)$ and is a function of airspeed (V_t) and altitude (h). Equation 5.1 along with the system output equation can be summarised as follows:

$$\dot{\mathbf{x}}(t) = \mathrm{f}(\mathbf{x}(t), \mathbf{u}(t)) \tag{5.7}$$

$$\mathbf{y}(t) = \mathrm{g}(\mathbf{x}(t), \mathbf{u}(t)) \tag{5.8}$$

where $\mathbf{x}$ is the state vector $[u\ w\ q\ \theta\ P_N\ h\,]^T$, $\mathbf{u}$ is the control input vector $[\eta\ \tau]^T$ and $\mathbf{y}$ is the user-defined output vector. Together (5.7) and (5.8) describe the longitudinal motion of the UAV.

5.3.2 Longitudinal Trim

Initially, all longitudinal flight tests assume steady state (non-accelerating), trimmed and rectilinear flight. Therefore all initial translational and rotational accelerations are zero ($\dot{u}, \dot{w}, \dot{q} = 0$), the initial sum of the forces and moments acting on the aircraft are zero (i.e. $X = Z = M = 0$) and the initial angular velocities are also zero ($q = 0$). The remaining flight variables at trim condition, for all flight tests, are as defined in Table 5.2.

Table 5.2 UAV model trim condition

α(deg)	γ(deg)	V_t (m/s)	h (m)	τ	η (deg)
7.55	0	32	1000	0.20	-12.98

5.3.3 The Unknown Inputs

To test the robustness of the SFDA schemes to unknown inputs we must be able to incorporate such inputs in the UAV model. Three types of unknown inputs are considered:

- Wind gust disturbances (system noise)
- Measurement (sensor) noise
- Parameter uncertainty

Wind gust disturbances are modelled as zero mean white, Gaussian gust disturbances acting on the angle of attack and angle of sideslip ($\alpha_{gust}, \beta_{gust}$). The statistical properties of the gust disturbances are defined in Table 5.3. The wind disturbance model presented here is not intended to accurately model the wind disturbances experienced during UAV flight. For example in our model we assume that α_{gust} is always in the direction of the aircraft's angle of attack orientation, however instead it may be applied at an angle and so we would need to consider the x-y components of α_{gust}. However we are more interested in analyzing the robustness of the SFDA schemes in the presence of a disturbance with known statistical properties, than accurately modelling the disturbance. Similarly we introduce measurement noise in the output parameters of the UAV model to simulate noise-corrupted sensors. Measurement noise is modelled as zero mean, white, Gaussian noise with statistical properties shown in Table 5.4.

Note that gyro sensor noise is *not* considered here. This is important as it simplifies the analysis of the EKF-based and NN-based SFDA schemes. If we were to add a large amount of noise on the pitch gyro, then the SFDA schemes will be highly susceptible to false alarms regardless of the accuracy of the EKF or NN pitch rate estimations. On the other hand including sensor noise in the parameters (as in Table 5.4) which are used as inputs to the EKF and NN is important, as it allows us to investigate the accuracy of the estimations under noise-corrupted inputs. Parameter uncertainties are also considered in the EKF equations but not in the NN, as the NN does not rely on a mathematical description of the system. Table 5.5 shows the parameter uncertainties which will have corresponding effects on the EKF system matrices (section 5.4).

Table 5.3 System noise standard deviations

Parameter	Standard deviation
α_{gust}	0.10 deg
β_{gust}	0.10 deg

Table 5.4 Measurement (sensor) noise standard deviations

Parameter	Standard deviation
V_t	0.42 m/s
a_x, a_y, a_z	0.07 m/s^2
α	0.03 deg
$\dot{u}, \dot{v}, \dot{w}$	0.03 m/s^2

Table 5.5 Parameter uncertainties

Parameter	Description	Uncertainty from nominal
m	Aircraft mass (kg)	±10%
x_{cg}	Aircraft centre of gravity (m)	±5%
I_y	Pitch inertia (kgm^2)	±10%
$C_{L\alpha}$	Aircraft lift curve slope (rad^{-1})	±20%
$C_{L\eta}$	Elevator lift effectiveness (rad^{-1})	±20%

5.4 Extended Kalman Filter (EKF)

The KF was first proposed by R.E. Kalman in 1960 [76]. Since then the KF has seen an overwhelming variety of applications [137]. Popular applications include the navigation field, where measurements from a global positioning system (GPS) and an inertial navigation system (INS) are combined into a KF measurement error model to provide an improved measurement. Another common example is their use in control applications as linear state estimators, where a KF is used to estimate otherwise immeasurable parameters required in control feedback loops [99]. KFs have also been widely applied in the FDI field where either the KF state estimations are directly used to generate a fault residual [78], or the statistical properties (e.g. mean) of the KF innovation sequence (i.e. difference between sensor measurements and KF estimates) are analysed for any unexpected deviations caused by the faults [81]. The KF is essentially a set of linear equations which are designed to produce optimum estimates of a vector of parameters by minimising the mean-squared estimation error at each iteration. In addition to the different KF applications, there have been a large variety of KF designs proposed over the years [139]. However the extended KF (EKF) remains one of the most popular approaches for nonlinear applications [138]. For a general introduction to KFs the reader is referred to [137-139].

The classical EKF relies on the discrete form of the system equations. Re-writing the UAV longitudinal motion equations (5.7)-(5.8) in discrete form and including the system and measurement noise (section 5.3.3), we can define the following (excluding the navigation equations):

$$\mathbf{x_k} = \mathrm{f}(\mathbf{x_{k-1}}, \mathbf{u_{k-1}}, \omega_{k-1}) \tag{5.9}$$

$$\mathbf{y_k} = \mathrm{h}(\mathbf{x_k}) + \mathbf{v_k} \tag{5.10}$$

where k is the sampling instant, $\mathbf{x} = [u\ w\ q\ \theta\ ]^T$, $\mathbf{u} = [\eta\ \tau]^T$, $\omega = \alpha_{gust}$ and $\mathbf{v}$ is the measurement noise vector. The system noise (ω) and measurement noise vector $(\mathbf{v})$ are assumed to be uncorrelated and are as defined in section 5.3.3 with variance Q and covariance structure **R** respectively; Q = 0.01 deg (Table 5.3) and **R** is a diagonal matrix with variances from Table 5.4 forming the diagonal elements. Additionally it is assumed that the system and measurement noise are stationary, i.e. they have constant variances.

The EKF is essentially a discrete Kalman filter which is continuously linearised about its estimated trajectories, i.e. state estimates $\mathbf{\hat{x}_k}$ [139]. The standard EKF equations are described in Fig 5.4 where the following Jacobians are calculated at each time step:

$$\mathbf{A}_k = \left.\frac{\partial \mathrm{f}}{\partial \mathbf{x}}\right|_{\mathbf{\hat{x}_{k-1}}, \mathbf{u_{k-1}}, 0}$$

$$\mathbf{G}_k = \left.\frac{\partial \mathrm{f}}{\partial \omega}\right|_{\mathbf{\hat{x}_{k-1}}, \mathbf{u_{k-1}}, 0}$$

$$\mathbf{C}_k = \left.\frac{\partial \mathrm{h}}{\partial \mathbf{x}}\right|_{\mathbf{\hat{x}_k^-}, 0}$$

where:

- $\mathbf{\hat{x}_k^-}$ is the *predicted* (*a priori*) state estimate
- $\mathbf{\hat{x}_k}$ is the *corrected* (*a posteriori*) state estimate
- $\mathbf{P_k}$ is the state estimation error covariance matrix. The diagonals of $\mathbf{P_k}$ are the mean-squared state estimation $(\mathbf{\hat{x}_k})$ error.
- $\mathbf{K_k}$ is the Kalman gain which is designed so as to minimise the diagonal elements of $\mathbf{P_k}$.

From Fig 5.4 we can see that at each iteration, the EKF executes two main steps; a prediction and a correction step. Firstly the states are predicted by using the available state equation (5.9), but setting the system noise (ω) to zero (as it can be difficult to predict). This gives the *a priori* state estimate $\mathbf{\hat{x}_k^-}$. Secondly this estimate is corrected by using the available sensor measurements $\mathbf{y_k}$. The final outcome is the *a posteriori* state estimate $\mathbf{\hat{x}_k}$. Therefore in our case, as we are interested in estimating the pitch rate, $\hat{q}_{EKF}$ is simply the third element in the state vector $\mathbf{\hat{x}_k} = [\hat{u}\ \hat{w}\ \hat{q}\ \hat{\theta}]$. In section 5.6 we define the input/output structure of the EKF used in the SFDA scheme.

If the state vector $\mathbf{x_k}$ in (5.9) includes angle of attack (α), then we can simply define vector $\mathbf{G_k}$ from matrix $\mathbf{A_k}$ instead of calculating $\mathbf{G_k}$ separately at each iteration. Ignoring the sampling instants for now, we know that for small perturbations about some operating point $[\mathbf{x}_0, \mathbf{u}_0]$, α can be approximated as $\alpha \approx w/V_t$ where w is the normal velocity and V_t is the total airspeed [140]. Therefore the state vector in (5.9) can be approximated as $\mathbf{x} = [u\ (\alpha V_t)\ q\ \theta\ ]^T$. If we apply appropriate scalar changes to matrix **A** the state vector can now be written as $\mathbf{x} = [u\ \alpha\ q\ \theta\ ]^T$; for example if the first row of **A** includes the aerodynamic derivatives $[\mathrm{x_u}\ \mathrm{x_w}\ \mathrm{x_q}\ \mathrm{x_\theta}\ ]$ then this row would now become $[\mathrm{x_u}\ (\mathrm{x_w} V_t)\ \mathrm{x_q}\ \mathrm{x_\theta}\ ]$. Once the appropriate changes have been made to matrix **A**, **G** is simply the second column vector of **A**, i.e. the column where α is seen to influence the state vector **x**.

It is well-known that the EKF estimations can diverge especially if the initial state vector estimate ($\hat{\mathbf{x}}_0$, defined by the user) is inaccurate and/or the sensor measurements are significantly contaminated with noise [139]. This is because the EKF matrices, $\mathbf{A}_k$, $\mathbf{C}_k$ and $\mathbf{G}_k$ are continuously updated based on the filter's state estimations and therefore inaccurate state estimates have a cumulative effect on filter performance. On the other hand, the KF has a fixed set of system equations and is therefore less susceptible to divergence problems. Consequently, it is common practice to first apply a KF and only if performance is poor, should a EKF be used. Additionally, it is important that the KF and EKF are appropriately tuned before application. This stage is known as filter tuning and is necessary regardless of whether there is a theoretical rationale for setting the variances Q and **R**, and the initial conditions $\hat{\mathbf{x}}_0, \mathbf{P}_0$. Filter tuning is applied until accurate state estimations ($\hat{\mathbf{x}}_k$) as well as filter convergence are achieved, i.e. the state estimation error variances (diagonal of $\mathbf{P}_k$) have settled to almost constant values.

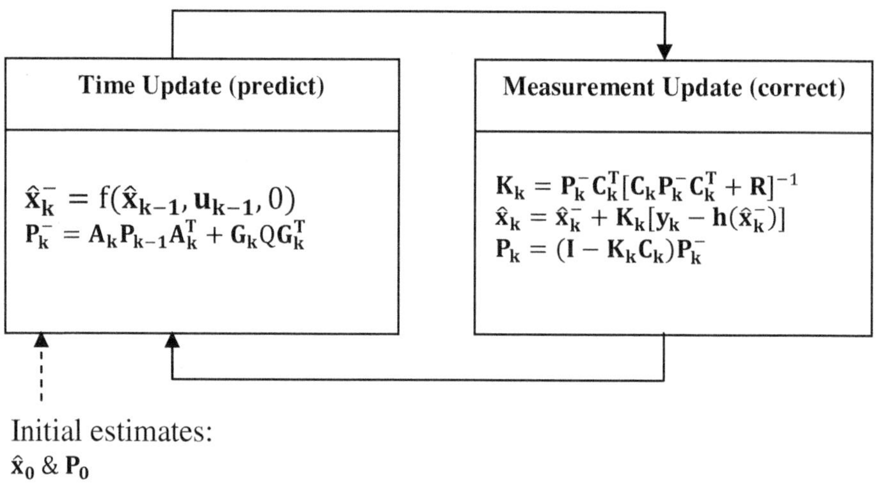

Fig. 5.4 Classical EKF equations [141]

5.5 Residual Structures

5.5.1 Residual Generation and Evaluation (RGE)

A residual is a fault indicator which is based on a deviation between the real sensor measurements and its model estimates. In its simplest form, a residual is generally the squared difference between sensor measurement (y_{real}) and model estimate ($\hat{y}$) [142]:

$$\bar{r}_k = (y_{realk} - \hat{y}_k)^2 \tag{5.11}$$

where $\bar{r}$ is the residual at sample instant k. Ideally the model estimates and sensor measurement are equal ($\therefore \bar{r} = 0$) and only in the event of a fault is $\bar{r} \neq 0$. In this case a threshold close to zero is suitable for fault detection. However due to the presence of residual noise (which is caused by the unknown inputs, section 5.3.3) the residual will be non-zero even when no faults are present, i.e. $\hat{y} \neq y_{real}$ and $\therefore$ $\bar{r} \neq 0$. This can result in false alarms and a simple solution is to raise the threshold. However this also increases the risk of not detecting the faults. Alternatively we can improve the basic residual generator in (5.11) so that residual noise is sufficiently dampened.

This is quite commonly done by computing the moving average of the residual in (5.11) [63]. In our case the arithmetic mean is chosen as the average and the weighted moving average of the past Ω residuals generated in (5.11), are calculated as follows (for $\bar{r}_j = 0, j \leq 0$):

$$r_{kRGE} = \frac{\varpi}{\Omega} \sum_{j=k-(\Omega-1)}^{k} \bar{r}_j \tag{5.12}$$

where r_{kRGE} is the residual generated at sample instant k and ϖ is the weight. The residual r_{kRGE} can then be evaluated against a pre-defined threshold for fault detection purposes. Overall this method of residual processing will be referred to as residual generation and evaluation (RGE).

5.5.2 Residual Generation, Padding and Evaluation (RGPE)

This method is designed to further reduce the effects of unknown inputs on the residual (i.e. residual noise) in comparison to the RGE approach so that false alarms and the number of undetected faults are reduced. It consists of the same steps as the RGE method however the residuals are now padded with *artificial* data. More specifically the Ω residuals ($\bar{r}$) in (5.12) are extended with padding data before the weighted moving average is calculated:

$$r_{kRGPE} = \frac{\varpi}{\Omega + p_{pad}} \left[\sum_{j=k-(\Omega-1)}^{k} \bar{r}_j + p_{pad} \min\{\bar{r}_{k-(\Omega-1):k}\} \right] \tag{5.13}$$

where r_{kRGPE} is the residual generated at sample instant k, p_{pad} is the number of padding points to be added and 'min' is the minimum function. Note that if padding is not applied then $p_{pad} = 0$ and $r_{kRGPE} = r_{kRGE}$ as expected.

Residual padding manages to extend the Ω samples in (5.12) with p_{pad} artificial data points **before** the weighted moving average is calculated. Each artificial data point added is equal to the **minimum** value in the original Ω samples. So for example if $\Omega = 4$ and the residual samples are $\bar{r}_{k-3:k} = \{1\ 0\ 3\ 4\}$, then these would be extended to $\{1\ 0\ 3\ 4\ \underline{0}\ \underline{0}\}$ where two extra zeros ($p_{pad} = 2$) are added. This extended residual set can then be used to calculate the residual in (5.13) where $r_{kRGPE} = \frac{\varpi}{4+2}[1+0+3+4+2(\underline{0})] = \frac{8}{6}\varpi$. (Note that the line under the zeros is only to clarify which data points are padding data points).

Residual padding is designed to reduce the overall average of each residual set so that the effects of unknown inputs are minimised. The 'min' function is used instead of simply adding 'zeros' because the latter would damp all the residuals (including ones caused by faults) while the former is more specific to the individual residual sets.

Consider this simple example to demonstrate the benefits of residual padding. Referring to (5.11)-(5.13) the following assumptions are made (no units are used):

1. $\Omega = 4$
2. $\bar{r}_{k-3:k} = \{0\ 0\ 0\ 0\}$ if there are no faults or unknown inputs
3. $\bar{r}_{k-3:k} = \{6\ 6\ 6\ 6\}$ if there is a fault present
4. $\bar{r}_{k-3:k} = \{12\ 12\ 0\ 0\}$ if there are unknown inputs present.
5. Only 2 padding points can be added, i.e. $p_{pad} = 2$
6. The weight ϖ is set to 1.
7. The residual threshold is set to 5.

In this example we have assumed that a fault has a continuous effect on the residual (assumption 3) while the unknown input has a spike-type effect and then settles to zero (assumption 4). This spike-type behaviour is quite common of residual noise. However in reality these spikes may be drifting and can eventually settle to a non-zero value.

Let us now consider three typical scenarios; Scenario 1 is when faults and unknown inputs are not present and therefore ideally the threshold is not exceeded, Scenario 2 is when only a fault is present and therefore the threshold should be exceeded, Scenario 3 is when only unknown inputs are present and therefore the threshold should not be exceeded (otherwise this results in a false alarm).

In Scenario 1, $\bar{r}_{k-3:k} = \{0\ 0\ 0\ 0\}$ and therefore using (5.12), $r_{kRGE} = 0$. Padding the group of residuals would give $\{0\ 0\ 0\ 0\ \underline{0}\ \underline{0}\}$ and so using (5.13) the residual r_{kRGPE} is also zero, $r_{kRGPE} = \frac{1}{4+2}[0+0+0+0+\underline{0}+\underline{0}] = 0$. Therefore in both cases (i.e. with or without padding) the threshold is not exceeded as desired.

In Scenario 2, $\bar{r}_{k-3:k} = \{6\ 6\ 6\ 6\}$ and $r_{kRGE} = 6$. Padding the data would give $\{6\ 6\ 6\ 6\ \underline{6}\ \underline{6}\}$ and so r_{kRGPE} is also 6. Therefore in both cases the threshold is exceeded and the fault is successfully detected.

The final scenario is when unknown inputs are present. $\bar{r}_{k-3:k} = \{12\ 12\ 0\ 0\}$ and so $r_{kRGE} = 6$ which exceeds the threshold resulting in an undesired false alarm. However if we extend the residual with padding data we get the set $\{12\ 12\ 0\ 0\ \underline{0}\ \underline{0}\}$. r_{kRGPE} is now 4 and so the false alarm is successfully avoided. Therefore residual padding has managed to damp the effects of the residual noise so that false alarms are avoided, but at the same time it avoided damping the fault effects on the residual (Scenario 2) so that the faults are detected.

This is obviously a simple example with no consideration to scenarios such as drifting faults, large residual noise spikes, or long-lasting unknown input patterns (i.e. consecutive noise patterns which do not instantly settle to zero). However the underlying concepts are the same. Residual padding attempts to damp the effects of unknown inputs on the residual. The assumption is that residual noise has a fast spike-type effect and then settles to a close-to-zero value. Therefore the artificial padding data (using the minimum function) added to the original residual set also has a close-to-zero value and so the overall residual average is reduced, i.e. damped. On the other hand it is assumed that faults (e.g. constant bias faults) have permanent effects on the residual, i.e. the residual does not settle back to a close-to-zero once the fault is introduced. Therefore the padding data added would not be close-to-zero (especially once the fault has reached its maximum value) and so the residual average may be reduced but not significantly. For example, a slow-drift fault will continue to increase and once it reaches its maximum value it will remain at that value and will not drop to zero (or a small value). However the opposite of the latter scenario are intermittent failures which occur on a periodic basis. If not careful residual padding may treat such failures as noise and therefore damp their effects on the residual. Intermittent failures are not considered in the SFDA tests carried out in this book. However it is important to test the sensitivity of the RGPE to intermittent failures if it is to be applied in a real system (see Chapter 8).

If residual noise can be sufficiently damped then it is also possible to amplify the overall residual (e.g. by selecting a larger weight value ϖ) so that small and slow-drift faults are detected, without running the risk of increasing the false alarm rate. However it is important that excessive padding and amplification are avoided as this can increase the fault detection time (or even completely damp the faults) and increase the false alarm rate respectively. Therefore the RGPE parameters as well as the RGE (section 5.5.1) parameters must be carefully tuned. The tuning parameters include, the residual threshold, p_{pad}, ϖ and Ω. The RGE and RGPE structures used in the SFDA tests are defined in Table 5.6.

In conclusion, residual padding is an algorithm designed to reduce the overall average of a residual set when residual noise is present. This in turn reduces the chances of the residual exceeding the threshold and resulting in false alarms. Residual padding reduces the average by extending the residual set with artificial data which is equal to the minimum data value in the residual set. The average of this extended residual set can then be taken. So far, residual padding assumes that residual noise has a spike-type effect and then settles to zero. The minimum value is therefore always close to zero and so padding the residual with zeros will reduce the overall average and the chances of false alarms. However for a more complete analysis, further work (as will be discussed later) will have to consider the performance of residual noise, when; the minimum value due to residual noise is not close-to-zero, faults also have the same pattern as residual noise (i.e. spike-type effects, also known as intermittent failures).

5.6 NN and EKF Input/Output Structure

The NN and EKF outputs are the estimated pitch rates $\hat{q}_{NN}$ and $\hat{q}_{EKF}$ respectively, which should each follow q_{real} and q_{ideal} in the non-faulty and faulty case respectively (section 5.1). We have not yet defined which flight data variables are used in the NN and EKF input set. Note that the input set of the EKF includes the parameters in the control vector **u** and the sensor measurement vector **y** (Fig 5.4, section 5.4). As we are only considering longitudinal motion, only the longitudinal flight variables are to be used in the NN and EKF input set. As discussed in section 5.3.1, aircraft longitudinal motion can be fully described by the axial force X, normal force Z and pitching moment M equations. These can be derived from Newton's second law of motion as follows [140]:

$$ma_x = X \tag{5.14}$$

$$ma_z = Z \tag{5.15}$$

$$I_y\dot{q} = M \tag{5.16}$$

The individual contributions to X, Z and M were defined in (5.2)-(5.4) and so if we substitute them into (5.14)-(5.16) we can define the full set of longitudinal variables describing the UAV longitudinal motion:

- Airspeed, V_t
- Angle of attack, α
- Altitude, h
- Pitch angle, θ
- Pitch rate, q
- Angular acceleration, $\dot{q}$
- Translational (normal) acceleration, $\dot{w}$
- Normal acceleration, a_z
- Axial acceleration, a_x

To decide which of the above variables are sufficient to give accurate estimates $\hat{q}_{NN}$ and $\hat{q}_{EKF}$, three criteria must be considered:

1. Pitch rate (as well as $\dot{q}$) must not be included in the input sets of the NN and EKF
2. The input set of the NN and EKF must be the same
3. The input set chosen depends on the EKF and not the NN structure.

Criterion 1 is important for two reasons (see for example Fig 5.1 where $y1_{real}$ is not included in the input set). The first reason is that the NN and EKF estimates are more likely to closely follow the sensor measurements q_{real} **even** if it is faulty, i.e. $\hat{q}_{NN} \cong q_{real}$ and $\hat{q}_{EKF} \cong q_{real}$. This can make fault detection difficult as the residual would be close to zero even if a fault is present in Sensor-q. Secondly when Sensor-q is faulty, its measurements may quickly contaminate the NN structure before training can be switched off and this can degrade the fault accommodation properties of the NN. Criterion 2 is necessary for fair comparisons of the NN and EKF. Note that the input set to the NN is chosen to be the same as the parameters in the EKF measurement vector **y** and not the parameters in control input vector **u** because the latter forms a necessary, pre-defined part of the EKF system equations while the former can be tuned by the user i.e. a set of EKF models applied to the **same** system model, must have the same control input vector **u** but their different measurement vectors **y** is what differentiates their performances. Criterion 3 is not so obvious but is important if the EKF is to produce accurate estimations. The reason for this is that the NN is only required to estimate the pitch rate. On the other hand, the EKF must estimate the entire state vector $\mathbf{x_k} = [u\ \alpha\ q\ \theta]^T$ accurately as the EKF updates its system equations (i.e. matrices **A**, **G** and **C**, Fig 5.4) based on the state estimate, $\hat{\mathbf{x}}_\mathbf{k}$. Therefore if we base the choice of the input set on the NN structure then the EKF may give poor estimations of the state vector $\mathbf{x_k}$. (Note that if a KF is used instead of the EKF then the inaccurate estimation of some parameters in the state vector may be tolerated as the KF has fixed system equations).

In conclusion the following steps are carried out to define the relevant longitudinal flight variables to be used in the NN and EKF (to estimate the pitch rate):

1. Define the full set of longitudinal variables (see above)
2. Attempt different combinations in the **y** matrix of the EKF.
3. The combination which results in the lowest EKF pitch rate estimation error is chosen as the final set which will be used in the EKF and NN.

The resulting NN and EKF input/output structures are as follows:

$$\hat{q}_{NNk} = NN(\alpha_{k-1}, a_{zk-1}, \dot{w}_{k-1}, V_{tk-1}) \tag{5.17}$$

$$\hat{q}_{EKFk} = EKF(\alpha_k, a_{zk}, \dot{w}_k, V_{tk}, \eta_{k-1}, \tau_{k-1}) \tag{5.18}$$

Note that the inputs to the EKF consist of the vector parameters $\mathbf{y_k}$ at sample instant k and the control input vector $\mathbf{u_{k-1}}$ at instant k-1 (Fig 5.4). The final NN and EKF input/output structures are shown in Fig 5.5 (note that the subscript '*real*' for the sensor measurements is omitted for simplicity purposes) and are also defined in Table 5.6.

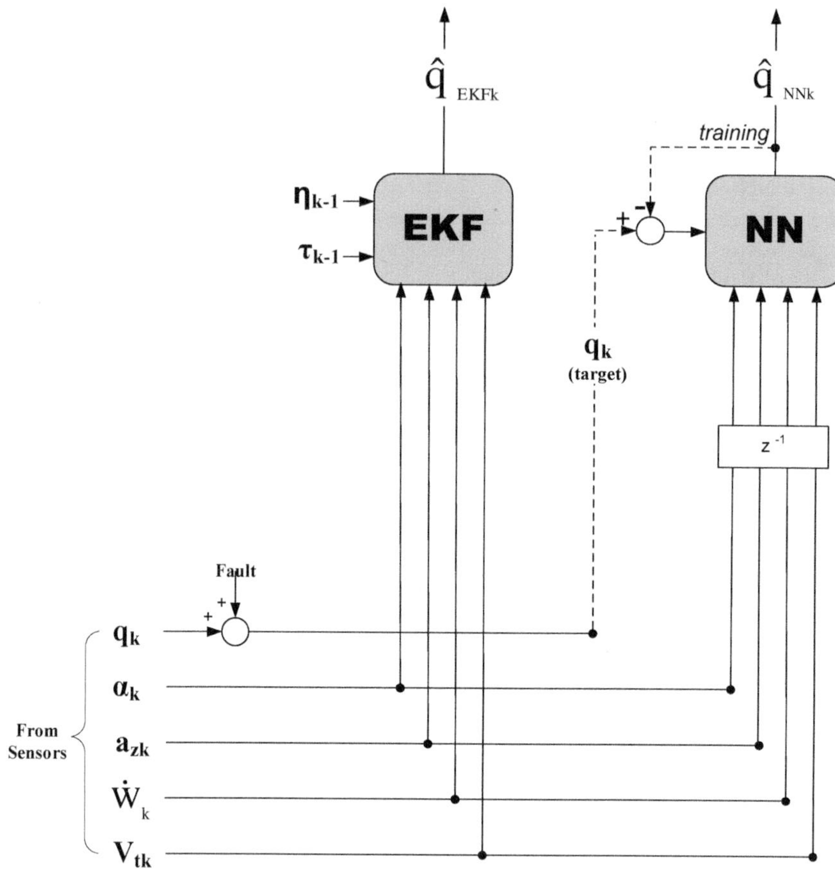

Fig. 5.5 EKF and NN input/output structure. Dotted lines indicate parameters required for NN training. Sensor faults can be present only in Sensor-q.

5.7 Sensor Fault Types

The sensor fault types considered in the SFDA tests are as follows:

- Constant bias faults: The sensor output gets stuck and outputs a constant bias.
- Additive faults: Additive faults are very common. A term is added to the normal sensor value as a result of temperature changes or calibration problems.

Additive faults are described by ramp functions (drift) and can be of step, soft or hard nature depending on the duration of the ramp duration T_R. T_R is approximately 0s, 1s and 4s for step-type, hard-additive and soft-additive faults respectively (Fig 5.6) [143]. An additive fault can be defined as follows [24]:

$$F(t)=\begin{cases}0 & t<t_{fault}\\ A(t-t_{fault})/T_R & t_{fault}<t<t_{fault}+T_R\\ A & t\geq t_{fault}+T_R\end{cases} \tag{5.19}$$

where the fault is introduced at t_{fault} and A is the fault magnitude. Incipient faults (small and/or slow drift faults) have a small A and/or a large ramp duration T_R. For large and small faults in Sensor-q it assumed that A = 2.4 deg/s and A = 1.2 deg/s respectively. The full list of fault types considered in the SFDA tests will be outlined in section 5.8.2.

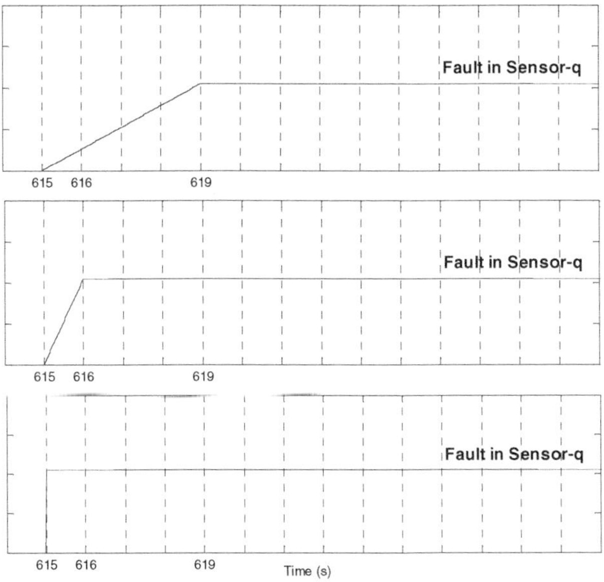

Fig. 5.6 Additive faults in Sensor-q (ignore fault magnitude). Top plot, middle plot and bottom plot are soft additive (T_R = 4s), hard additive (T_R = 1s) and step type ($T_R \cong$ 0s) faults respectively.

5.8 SFDA Application to UAV Model

5.8.1 NN Training

The NN structure chosen in the SFDA tests is the EMRAN RBF NN discussed in Chapter 4. Prior to any SFDA test, the NN must be initialised to avoid poor initial estimations. This stage is referred to as NN offline training (Chapter 4, section 4.2). Note that as this stage is only to initialise the NN structure, no fault scenarios are considered in the training data. Once the NN offline training stopping criteria is achieved, the NN structure is frozen and used in the SFDA tests where it can then continue training (online training). Note that the offline and online training algorithms are the same (Chapter 4, section 4.1.3) but will have different tuned parameters (e.g. learning rates). (225s of flight data is used during the offline training stage).

5.8.2 SFDA Test Outline

Four SFDA schemes are designed and compared. These include NN-RGE, EKF-RGE, NN-RGE and EKF-RGPE. The NN and EKF input/output structures chosen are as discussed in section 5.6 and the SFDA tests are carried out on the UAV model presented in section 5.3. The SFDA tests consider different levels of unknown inputs (section 5.3.3) and faulty types (section 5.7) and in each test, Sensor-q is only allowed to fail **once**. Overall there are 244 separate SFDA tests (800s of flight data each) which include; **7 unknown input configurations, 8 fault types and 4 SFDA schemes**. Note that the NN is initialised (via offline training) only **once** using the Configuration 1 settings (to be defined).

The 7 configurations considered are as follows:

- **Configuration 1:** No unknown inputs, i.e. the system and measurement noise standard deviations are as in Tables 5.3 and 5.4 with correct Q and **R** variance structures in the EKF and no parameter uncertainties.
- **Configuration 2:** Parameters in Table 5.5; mass, x_{cg}, I_y, $C_{L\alpha}$, $C_{L\eta}$ are underestimated by 5%, 2.5%, 5%, 10%, 10% in the EKF equations.
- **Configuration 3:** System and measurement noise standard deviations (Tables 5.3 and 5.4) are doubled in the UAV model.
- **Configuration 4:** Configurations 2 and 3 are combined.
- **Configuration 5:** Same as Configuration 2 but underestimation of 10%, 5%, 10%, 20%, 20%.
- **Configuration 6:** Same as Configuration 3 but 10 times larger standard deviations.
- **Configuration 7:** Configurations 5 and 6 combined.

The reasons for choosing these configuration settings are as follows:

- Configuration 1: To investigate the general modelling and SFDA performance capabilities when the EKF system equations are perfectly modelled and when the NN is initialised using the same data statistical properties (i.e. same system and measurement noise standard deviations) used during offline training.
- Configuration 2: To investigate the robustness of the EKF to parameter uncertainties. As will be discussed in Chapter 8, future work must also investigate the performance of the NN when parameter uncertainties are present in the UAV model.
- Configuration 3: To investigate the robustness of the NN and EKF to an increase in system and measurement noise. Therefore we consider the scenario when the NN is initialised (via offline training) using one set of data (i.e. with Configuration 1 settings), but then during application the system and measurement noise increase. This configuration can also be seen as an

underestimation of the variances in the Q and **R** matrices of the EKF equations (section 5.4).

- Configuration 4: To simulate the real scenario when both parameter uncertainties are present in the EKF equations and the system and measurement noise levels have increased.
- Configurations 5, 6 and 7: These are repetitions of Configurations 2, 3 and 4 respectively except that the unknown inputs have increased.

It is important to note that parameter uncertainties are only associated with the EKF equations and **not** in the UAV model. This is because we are mainly interested in investigating the robustness of the SFDA schemes, which are based on a mathematical description of the system, to parameter uncertainties. However future work must also consider the scenario when parameter uncertainties are present in the UAV model, and test the NN performance in such scenarios (discussed in Chapter 8). On the other hand, changes in the system and measurement noise levels are considered in the UAV model (Configurations 3 and 6). This is to demonstrate the real scenario where the statistical properties of noise are generally unknown and time-varying. For this reason, NN offline training is implemented only once, using the Configuration 1 settings, and this initial structure is then used in all SFDA tests. In other words we consider the scenario when the NN is developed using one set of data but then during application the noise statistics changes. In theory, unlike the EKF, the NN should be able to adapt (via online training) to the time-varying system.

The 8 fault types considered are as follows (refer to section 5.7):

- **CB_L:** Constant bias large (A=2.4 deg/s)
- **ST_L:** Step-type large (A=2.4 deg/s)
- **HA_L:** Hard additive large (A=2.4 deg/s,T_R=1s).
- **SA_L:** Soft additive large (A=2.4 deg/s, T_R=4s)
- **CB_S:** Constant bias small (A=1.2 deg/s)
- **ST_S:** Step-type small (A=1.2 deg/s)
- **HA_S:** Hard additive small (A=1.2 deg/s, T_R=1s)
- **SA_S:** Soft additive small (A=1.2 deg/s, T_R=4s)

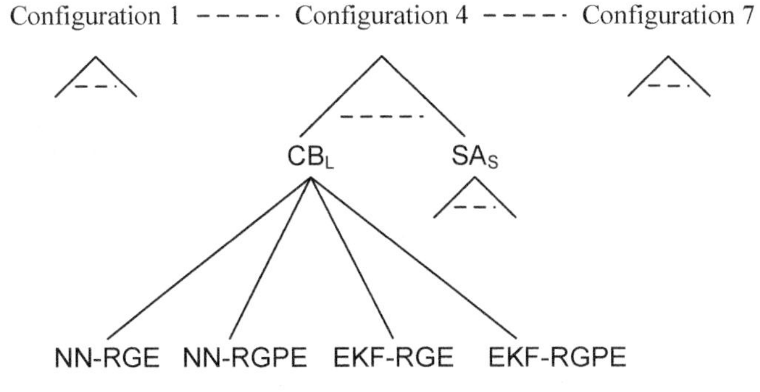

Fig. 5.7 Tree-diagram showing the different SFDA tests. 7 configurations, 8 fault types and 4 different SFDA schemes. Overall there are 244 separate tests.

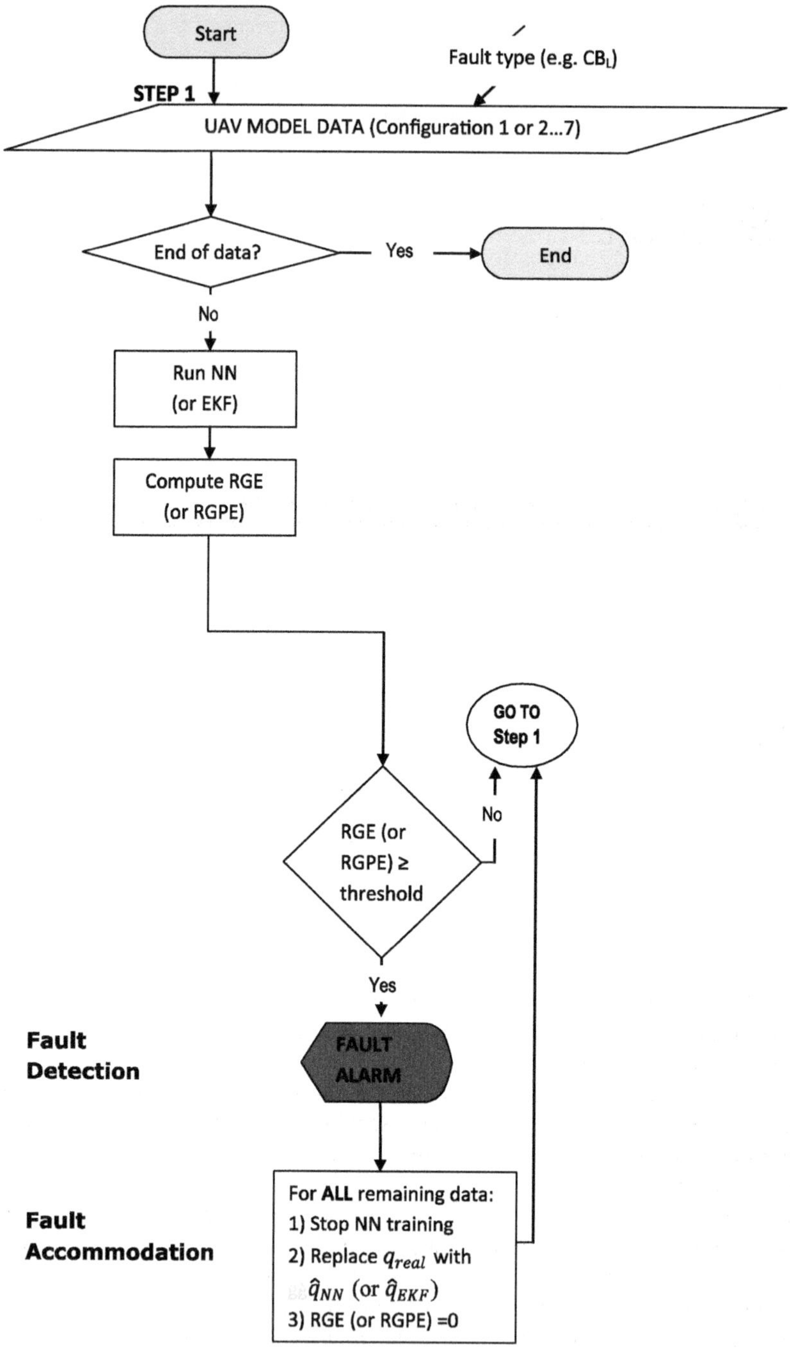

Fig. 5.8 Flow chart for one SFDA test (Note: one pass through the flowchart per sample instant, *k*)

Fig 5.7 shows a tree-diagram of the different SFDA tests where each full branch is 1 SFDA test. For example one of the branches is the SFDA test which uses Configuration 4 data settings and the NN-RGE scheme to detect a CB_L fault.

Fig 5.8 shows the flowchart for the SFDA tests where 1 pass through the flow chart (i.e. from Start to End) represents 1 SFDA test. Note that as we are assuming that the sensor can only fail once, the fault alarm is triggered only once, which is why we set the residuals (RGE and RGPE) to zero once a fault is detected (see last section of the flowchart).

5.8.3 SFDA Performance Indicators

To assess the performances of the SFDA schemes, two indicators are calculated for each test (Fig 5.9):

$$FA = \frac{t_{fa}}{t_{failure}} \text{x}100\,\% \tag{5.20}$$

$$DR = \frac{r_{af}}{r_{pr}} \tag{5.21}$$

FA is a false alarm indicator where t_{fa} is the total time the residual remains above the threshold prior to fault detection, and $t_{failure}$ is the time at which the fault is generated. *FA* indicates the percentage of time false alarms are present before the fault is actually detected. On the other hand, *DR* is a fault sensitivity (detectability) indicator (Fig 5.9). It calculates the ratio between the main residual magnitude at fault detection (r_{af}) and the maximum residual before the fault is introduced (r_{pr}). Note that as the effect of a fault on the residual is of ramp-type, r_{af} is calculated as the maximum residual value reached after fault detection (i.e. after residual = threshold) and before it starts to settle or decrease. In our tests all faults are introduced at 615s. Additionally, the mean (absolute) estimation error

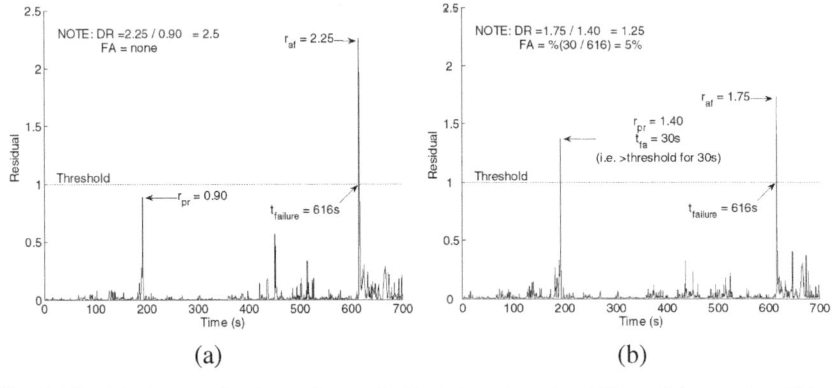

Fig. 5.9 Residual examples (no units used). Fault introduced at 615s and detected at 616s.

(*MEE*) and the variance of the estimation errors (*VEE*) are calculated wherever suitable. Note that prior to fault detection, $\hat{q}_{NN}$ and $\hat{q}_{EKF}$ are compared to q_{real}, and following fault detection (i.e. during the fault accommodation stage) the estimations are compared to q_{ideal} (section 5.1).

Results

The NN and EKF as well as the RGE and RGPE structures are outlined in detail in Table 5.6 and some of the UAV flight data is shown in Fig 5.10. Note that the NN input parameters are the same as the EKF output vector **y** (refer to section 5.6). Together Tables 5.6-5.12 and Figs 5.10-5.19 display the SFDA results. A careful analysis of the results reveals the following:

- **Table 5.7, Figs 5.11 and Fig 5.12:** These show the NN offline training results. Offline training is only associated with the NN which is why *MEE* for the EKF (and KF) is constant in Fig 5.11. It is only included for comparative reasons. The number of epochs elapsed before NN offline training is stopped is 2411 (1hr 25mins offline training time when implemented on a 1.6GHz Pentium processor). From Fig 5.11 it can be seen that the steep error gradients occur at the initial stages of training. This is expected as the EMRAN RBF NN starts with zero hidden units and gradually develops its structure with further training (Fig 5.12). The *MEE* decreases with more training epochs to a value close to the *MEE* of the EKF. On the other hand, the KF has a comparatively much larger *MEE*. Fig 5.12 shows the hidden neuron pattern for the EMRAN RBF NN during the offline training stage. The maximum number of hidden neurons achieved is 8, which is reached after almost 200 training epochs. Training beyond this point only consists of updating the free parameters (weights, centres, widths) but the number of hidden neurons will remain fixed.
- **Table 5.8:** This shows the estimation errors during the SFDA tests (when no faults are present). In general we notice that the *MEE* for the NN and EKF are close in value for Configurations 1-5. However for Configurations 6-7 the *MEE* increases significantly for the EKF but not for the NN. We also notice that the KF estimation performance is generally poor for all configurations.
- **Fig 5.13:** This plots the mean execution times for the first 100 data samples of the SFDA test (considered only for Configuration 1 settings and when no faults are introduced). The process is repeated for 100 iterations and the mean is then calculated which resulted in 1.90 ms, 0.69 ms and 0.21 ms mean execution times for the EKF, NN and KF respectively. (Note that the programs were run on a 1.6GHz Pentium processor). As expected, the KF has the fastest execution time as it consists of a fixed set of equations while the EKF system equations are continuously linearised at each time step and the NN structure must be updated via the training algorithm. It can also be seen that it takes less time to train the NN than to update the EKF equations. However in general the processing time per data sample for all 3 modelling techniques is lower than the flight data sampling time (0.02s).

- **Fig 5.14:** This plots $\hat{q}_{NN}$ and $\hat{q}_{EKF}$ for both Configurations 1 and 6. We only show the NN-RGE and EKF-RGE schemes as it is the method of modelling (EKF vs. NN) and not the residual structure which is important for the analysis carried out here. The corresponding estimation errors (i.e. $[q_{real} - \hat{q}_{NN}]$ & $[q_{real} - \hat{q}_{EKF}]$) for each estimation plot are also shown. In general we notice that for Configuration 6, where the system and measurement noise is maximum (section 5.8.2), the EKF-RGE estimation errors increase significantly (Fig 5.14d) in comparison to the NN-RGE (Fig 5.14c). This shows that 1) The EKF performance is poor when the system and measurement noise are incorrectly modelled in the EKF equations, 2) The NN online training algorithm has suitable adapted to the increase in system and measurement noise.
- **Tables 5.9-5.11:** These show the fault detection results for all SFDA tests, where Tables 5.10-5.11 and Table 5.9 are a breakdown of the results and a summary of the results respectively. Comparing NN-RGE and NN-RGPE we will notice, from Table 5.9, that the NN-RGE has 4 undetected faults (UDs) regardless of the configuration settings. These faults were found to be mainly the hard and soft additive faults; SA_L, ST_S, HA_S and SA_S. On the other hand the NN-RGPE detects such faults with almost no UDs. It also has on average a higher detectability ratio (*DR*). For example in Table 5.111, the NN-RGE does not detect the SA_S fault but the NN-RGPE detects the fault with a *DR* of 1.81. Comparing the fault types we will notice, from Tables 5.10 and 5.11, that **all** constant bias faults (CB_L and CB_S) are successfully detected, regardless of the configuration settings or SFDA scheme used. They are also detected much quicker (Table 5.10) and result in higher *DR* (Table 5.11) in comparison to other fault types. Comparing the NN based and EKF-based schemes we will notice from Table 5.9, that EKF-based schemes have fewer cases of undetected faults (i.e. lower UD). However they also have a higher percentage of false alarms (*FA*). So for example in Table 5.9 (Configuration 7), the EKF-RGE has a *FA* of 2.32% while the NN-RGE has a *FA* of only 0.09%. Comparing the NN-based and EKF-based schemes for the different configuration settings we will notice, from Table 5.9, that for Configurations 1-5, the EKF-RGPE has a much higher *DR* than the NN-RGPE. For example in Table 5.9 (Configuration 1), *DR*=104.72 for the EKF-RGPE but is nearly 3 times smaller (*DR*=40.23) for the NN-RGPE. On the other hand, for Configurations 6-7 the *DR* for the EKF-based schemes reduce significantly in comparison to the NN-based schemes. For example in Table 5.9 (Configuration 7), *DR*=0.06 and *DR*=23.40 for EKF-RGPE and NN-RGPE respectively. A *DR* of less than 1 shows that the residual peak due to a fault is actually **smaller** than the residual noise present prior to fault detection, i.e. $r_{af}>r_{pr}$ (section 5.8.3). This is one indication that the EKF estimation performance degrades for configurations with higher amounts of system and measurement noise. In summary, from Table 5.9, we will find that RGPE-based schemes increase the mean detection time (*MT*) by 120% but they also decrease the false alarms by almost 60% and have a 6 times greater detectability ratio than RGE-based schemes. Additionally in comparison to EKF-based schemes, NN-based schemes have a higher number of undetected faults but they also result in higher detectability rations and lower false alarm

rates for large amounts of system and measurement noise, i.e. Configurations 6 and 7.

- **Figs 5.15-5.17:** To demonstrate the results obtained in Table 5.9 more clearly, the residual plots (RGE and RGPE) for some of the tests are shown in Figs 5.15-5.17. Results for only HA_L (large hard additive) and SA_S (small soft additive) faults, and Configurations 1, 4 and 7 are shown. Note that in these figures the residual is not set to zero and NN training is not switched off if the fault is detected (which should be the case as discussed in section 5.1). However, this is only the case for Figs 5.15-5.17, so that the residual patterns are clearer. As we observed earlier in Table 5.9, we notice that the NN-RGE does not detect the SA_L fault (Figs 5.15c, 5.16c, 5.17c) while the NN-RGPE detects the fault successfully (Fig 5.15b, 5.16b, 5.17b). Furthermore for Configuration 7, the NN-RGE is unable to avoid false alarms (Figs 5.17a and 5.17c) in comparison to the NN-RGPE (Figs 5.17b and Fig 5.17d). We can of course raise the RGE threshold so that the false alarms are avoided, but this would also increase the number of undetected faults (UD). So for example in Fig 5.17a, we note that the peak when the fault is detected and the maximum peak before the fault is detected are very close in value, i.e. $r_{af} \cong r_{pr}$ (section 5.8.3). In fact the *DR* is less than 1 (*DR*= 0.76, Table 5.117). Therefore if we raise the threshold to avoid the false alarms we may consequently not detect the fault. If on the other hand we observe the same scenario in Fig 5.17b, we notice that the RGPE magnifies the difference between these peaks with a much higher *DR* of 4.59 (Table 5.117). This magnified difference allows us to amplify the residual sufficiently so that the fault is detected but at the same time the false alarms are avoided. On the other hand the EKF-based schemes successfully detect both HA_L and SA_L faults regardless of the configuration or residual structure used (Fig 5.15e-h, Fig 5.16e-h, Fig 5.17e-h). This was also seen in Table 5.9 where UD was generally zero for EKF-based schemes. However they are also more susceptible to false alarms in comparison to NN-based schemes especially for Configuration 7 (Fig 5.17e-h).
- **Table 5.12 and Fig 5.18:** These assess the NN fault accommodation properties. Fig 5.18 shows the NN fault accommodation results for large faults (CB_L, ST_L, HA_L and SA_L) while Table 5.12 summarises the results for all fault types and residual structures. From Fig 5.18a we notice that the large constant bias fault (CB_L) has resulted in poor NN fault accommodation performance in comparison to other fault types. This is despite the faster fault detection times found earlier in Table 5.10 for constant bias faults, i.e. despite the fact that we manage to detect the fault much quicker and therefore switch off NN learning before the NN structure is contaminated with faulty data. From Table 5.12 we notice that *MEEs* and *VEEs* for the NN are almost constant for all configurations while for the EKF they increase significantly for Configurations 6-7. For example in Table 5.12 the EKF-based schemes have a *VEE* of 3.28 for Configuration 7 while the *VEE* for the NN-based schemes is almost 30 times smaller (*VEE* = 0.11).

Discussion

In comparison to the EKF and NN we find that the KF shows poorer estimation performance (Table 5.8). It is well known that the KF equations are designed to give the minimum estimation error variance; however, this is only true if there are no modelling errors. For nonlinear applications, the KF will generally perform well within a certain operating range, i.e. close to the operating points about which the KF is linearised. From our tests it is clear that the EKF outperforms the KF with lower *MEEs* and *VEEs* but this is also at the cost of increasing the execution time (Fig 5.13). For one sample of data the KF reduces the execution time by nearly 89% in comparison to the EKF. Therefore a logical question would be whether the KF estimation errors can be tolerated with the benefit of increasing the execution time onboard the aircraft. To answer this let us consider the *MEE* for the KF in Table 5.8 (Configuration 1). The *MEE* is 1.28deg/s. In comparison to the small faults which have a magnitude of 1.2deg/s (A=1.2 deg/s, section 5.7), the *MEE* for the KF is actually higher. This can make it difficult for the SFDA scheme to differentiate between routine estimation errors and small magnitude faults. In SFDA applications the estimation errors must be as low as possible to avoid any false alarms and a reasonable guideline is that the *MEE* must be significantly lower than the smallest expected magnitude of a fault so that the SFDA scheme can be made sensitive to such faults. Therefore whilst having the lowest execution time, the KF is not suitable in our SFDA application. On the other hand the NN-based schemes and the EKF-based schemes may have larger execution times but they also have an average *MEE* of only 0.21 deg/s and 0.22 deg/s respectively.

The NN must be given sufficient offline training time before it can be used in the SFDA tests. This was seen in the offline training stage where the NN required 2411 training epochs before the stopping criterion was satisfied. There are no time limits on how long the NN should be trained offline as this stage is not done during the flight tests. Additionally in our case the NN was trained offline with randomly selected data sets, however in real implementation it can be more robust to train the NN with a larger data set, as well as data which covers the entire domain of validity of the flight tests, i.e. training data which covers the entire flight envelope.

In general it can be seen that NN-based schemes start to significantly outperform the EKF-based schemes for Configurations 6-7, i.e. when the system and measurement noise are significantly increased. This was seen in different parts of the results. Firstly, this was seen in Table 5.8 were the *MEE* for the EKF increased by almost 50% from Configuration 5 to Configuration 6-7 while the *MEE* for the NN remained almost constant. Secondly this was seen in Table 5.9, where the false alarm properties (*FA*) are higher for the EKF and the detectability ratio (*DR*) is lower in comparison to the NN-based schemes. Thirdly this was seen in Fig 5.14 where the estimation errors for the EKF increased significantly for Configuration 6 (Fig 5.14d) in comparison to the NN (Fig 5.14c). Finally this was also seen in the fault accommodation results (Table 5.12) where *MEE* and *VEE* increased significantly for the EKF but remained more or less constant for the NN-based schemes. Two important conclusions can therefore be drawn. One conclusion is that the EKF seems to be robust to small and large amounts of parameter uncertainties and robust to small amounts of system and measurement noise. Another conclusion is that, provided the NN structure has been suitably initialised during the offline training stage, the NN online training algorithm allows it to suitably adapt to time-varying systems, with better SFDA performances than the EKF as well as faster execution times (Fig 5.13).

In general, however, the EKF-based schemes outperform the NN-based schemes in terms of the fault detection properties (i.e. in terms of the fault detectability ratios and number of undetected faults). From Table 5.9 we will notice that the NN generally has smaller DRs (especially for the RGPE structures) than the EKF for Configurations 1-5 and a higher number of undetected faults (UD). In fact the NN-RGE did not detect soft additive faults and most of the small magnitude faults (i.e. incipient faults), regardless of the configuration settings. The reason for this is that the NN online training algorithm is not designed to differentiate between flight data which it should adapt from and faulty data which it should avoid. Consequently it can learn the fault and if this fault is not sufficiently large, the residual generated may not be high enough to exceed the threshold. This is clearly seen in Fig 5.15c where a peak at around 615s (i.e. around the time the fault is introduced) is visible but is not large enough to exceed the threshold. The EKF on the other hand does not suffer from this as we can see from Fig 5.15g where the fault is successfully detected. Therefore, the same property (online training algorithm) which makes the NN superior to the EKF is the same property which causes the NN-based SFDA schemes to have higher numbers of undetected faults and lower detectability ratios. However we can tune the online learning rate and more importantly improve the residual generator so that the incipient faults are detected. This was seen from the NN-RGPE results where the number of undetected faults was generally zero and the detectability ratio was increased in comparison to NN-RGE.

Let us now compare the different residual structures RGE and RGPE. In general we can see from Table 5.9 that the RGPE results in lower *FA* and UD than the RGE approach for all configurations. In fact in a separate test where Ω (i.e. averaging size) in (5.13) was increased and the median function was used instead of the arithmetic mean, the RGE approach still had higher FAs and UDs than the RGPE approach. Therefore RGPE manages to successfully damp the effects of unknown inputs on the residual and at the same time amplify the residual so that incipient faults are more easily detected. This is best seen in Fig 5.17c and Fig 5.17d where for the RGE approach the fault peak (at around 615s) is not large enough to exceed the threshold and false alarms are present, while for the RGPE approach the fault peak has been sufficiently amplified to exceed the threshold and false alarms are avoided.

The drawback of the RGPE approach is the increase in fault detection times. For example in Table 5.9 (Configuration 7), the mean detection time (*MT*) for the EKF-RGE is 1.74s and for the EKF-RGPE it increases by almost 191% to 5.06s. This is expected because residual padding (associated with the RGPE) essentially dampens the residual which can consequently delay fault detection if not carefully tuned. However in comparison to other work, the fault detection times remain within acceptable limits [24-26, 142, 143].

In conclusion, the benefits of the RGPE are the reduction of the false alarms rates and number of undetected faults but its drawback is the increase in fault detection time. Therefore for faults which are detected successfully by the RGE approach (such as constant bias faults) and in applications where false alarm rates are low, it is not necessary to use the RGPE method as it increases the fault detection time. Of course in reality one would not know *a priori* which sensor fault classes are more likely to occur and so a trade-off is required where sufficient amounts of padding and amplification can be chosen to detect the incipient faults and avoid false alarms, but at the same time high amounts of padding are avoided to keep the fault detection times within acceptable limits (generally dictated by the user). Similarly a suitable threshold must be selected based on experience. In our case a p_{pad} of 50 samples (1s), an amplification of 40 ($\varpi = 40$) and a threshold of 0.004 $(rad/s)^2$ were found most suitable (Table 5.6). As we discussed earlier, the choice of the NN learning rate is also crucial. In general the offline learning rate is set to be higher than the online learning rate. This is because a high online learning rate can cause the NN to learn the faulty measurements while in the offline training stage, a high learning rate is required to adequately build the NN structure. An offline and online learning rate of 0.04 and 0.0007 respectively, were found most suitable for our tests.

From the results we also found that constant bias faults are more easily detected than any other fault class. This is expected as the fault pattern is significantly different than the sensor measurements while additive faults simply result in a fixed offset from the sensor measurement. The question is which fault classes should be detected faster than others? Despite their drift-type effect, incipient faults can eventually severely damage the plant if left undetected for too long [9]. However a slow fault detection time for incipient faults can be tolerated. On the other hand large and constant bias faults must be quickly detected. There are two reasons for this. The obvious reason is that they can cause instant damage to plant performance especially if the sensor measurements are used in a control feedback loop. The second reason is that the NN structure can be significantly contaminated from the faulty measurements and consequently degrade the fault accommodation performance. This was seen in Fig 5.18. In Fig 5.18a the NN accommodated estimate (i.e. the NN estimate after the fault was detected) is poorer in comparison to other scenarios such as ST_L faults in Fig 5.18b. Therefore, the NN would require a fast fault detection times so that training is switched off before the NN structure is contaminated. This sensitive dependency on the fault detection time generally results in larger accommodation errors than the EKF as seen in Table 5.12. The effects of NN contamination will be clearer when multiple sensor faults are considered in Chapter 6, as in this case the NN accommodated estimate is feedback to other NN models.

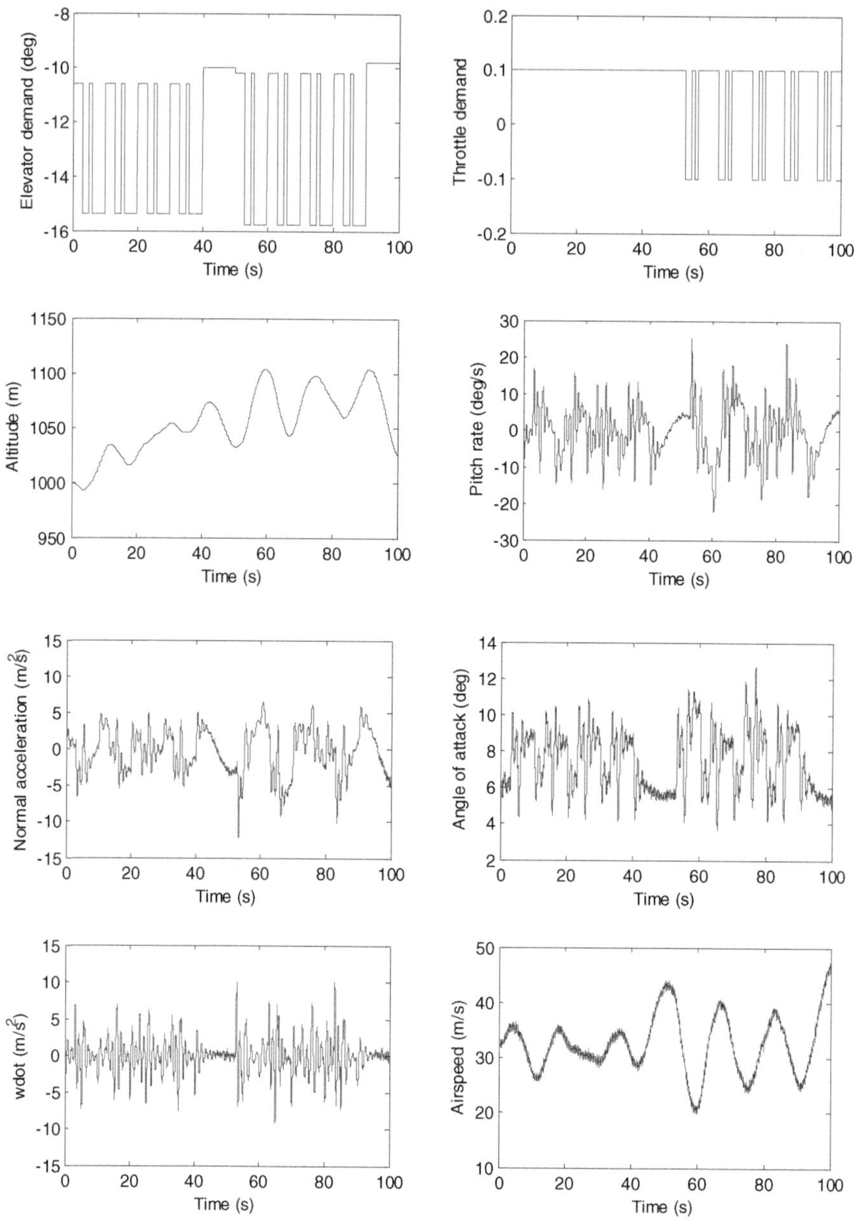

Fig. 5.10 (a) UAV model flight data for SFDA tests (Configuration 1, shown only for first 100s)

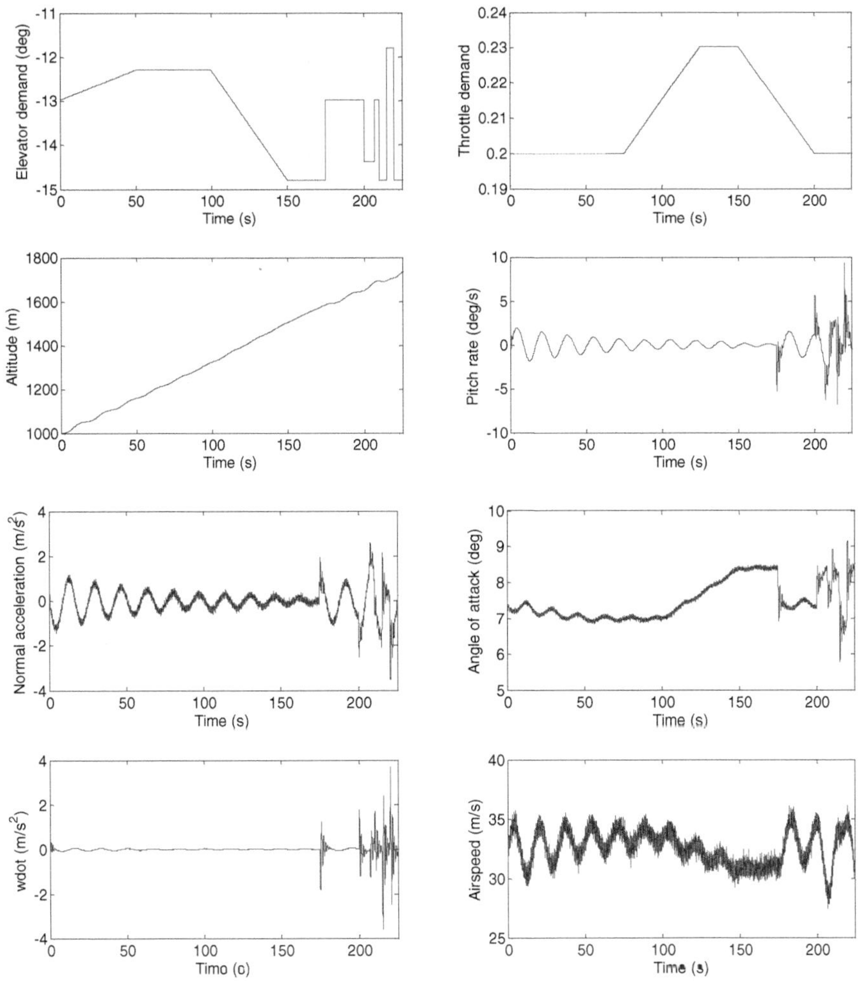

Fig. 5.10 (b) UAV model flight data for NN training (Configuration 1, 225s of flight data)

Table 5.6 Summary of NN, EKF structures and RGE, RGPE structures

NN & EKF STRUCTURE		
	NN	EKF
Input parameters:	$[\alpha\ a_z\ \dot{w}\ V_t]$	$[\alpha\ a_z\ \dot{w}\ V_t\ \eta\ \tau]^a$
Output parameter:	$\hat{q}_{NNk}$	$\hat{q}_{EKFk}$ [b]
No. of input neurons:	0	-
No. of hidden neurons (max):	9	-
No. of output neurons:	1	
Input data normalisation:	0-1	-
NN learning rate:	0.0007	-
$[E1\ \ E2\ \varepsilon_{max}\ \varepsilon_{min}\ \gamma_{dy}\ k_{op}]^c$:	[1e-4 0.01 0.6 0.3 0.997 1e-6]	-
RESIDUAL STRUCTURE		
	RGE	RGPE
Weight (ϖ) [d]:	1	40
Averaging size (Ω) [d]:	50	50
Padding points (p_{pad}) [d]:	-	50
Threshold:	0.001 $(\text{rad/s})^2$	0.004 $(\text{rad/s})^2$

-: Inapplicable
a: This is made up of the EKF output vector $\mathbf{y}_k$ and input vector $\mathbf{u}_{k-1}$ (Fig 5.4)
b: This is obtained from the EKF state vector estimate $\hat{\mathbf{x}}_k$
c: Refer to Chapter 4, Eq. (4.11)-(4.16).
d: Refer to Eq. (5.10)-(5.11)

Table 5.7 NN mean estimation error (*MEE* in deg/s) at the end of offline training for the test data set. EKF and KF errors also shown.

	NN		EKF	KF
	Epoch	*MEE*	*MEE*	*MEE*
Config. 1	2411	0.26	0.22	1.12

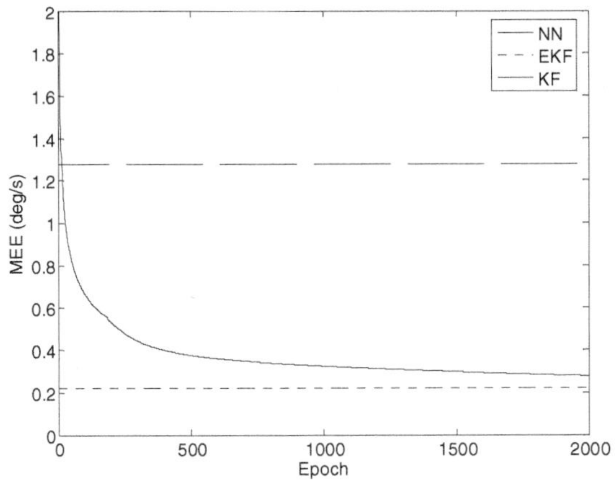

Fig. 5.11 NN offline training error history. (EKF and KF estimation errors also shown).

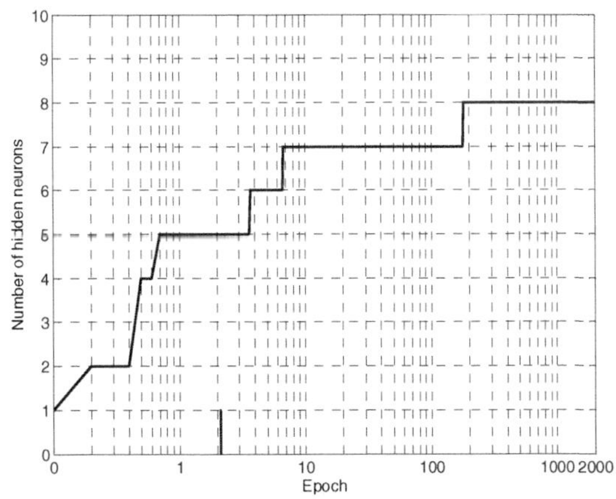

Fig. 5.12 NN hidden neuron pattern during offline training (shown only for first 2000 epochs)

Table 5.8 NN, EKF and KF mean estimation error (*MEE*, in deg/s) and estimation error variance (*VEE*, in deg^2/s^2) for the SFDA tests when no faults are present.

	NN		EKF		KF	
	MEE	*VEE*	*MEE*	*VEE*	*MEE*	*VEE*
Config. 1	0.21	0.06	0.22	0.12	1.28	5.24
Config. 2	0.21	0.06	0.23	0.13	1.28	5.24
Config. 3	0.21	0.05	0.29	0.22	1.29	5.21
Config. 4	0.21	0.05	0.30	0.23	1.29	5.21
Config. 5	0.21	0.06	0.31	0.76	1.28	5.24
Config. 6	0.22	0.06	0.63	1.13	1.31	4.99
Config. 7	0.22	0.06	0.61	1.35	1.31	4.99

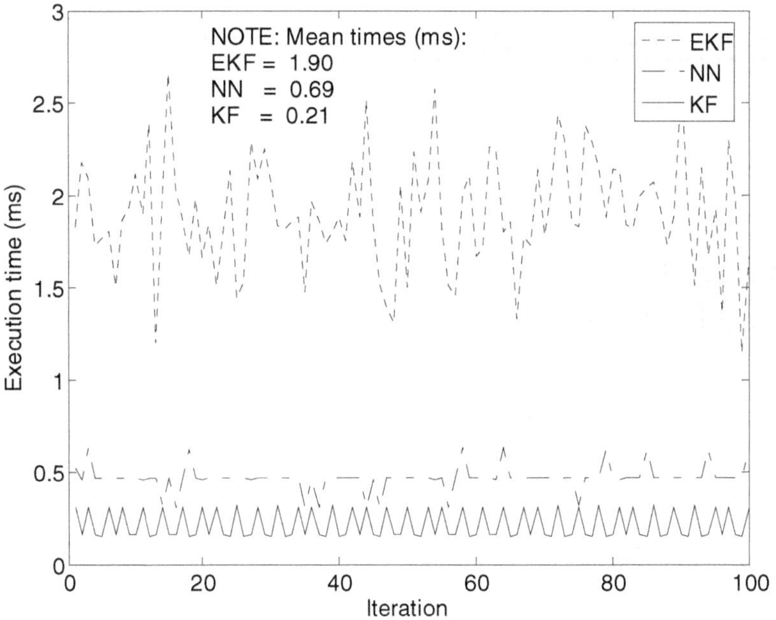

Fig. 5.13 Mean execution times per data sample. Mean calculated for first 100 data samples and then repeated for 100 iterations.

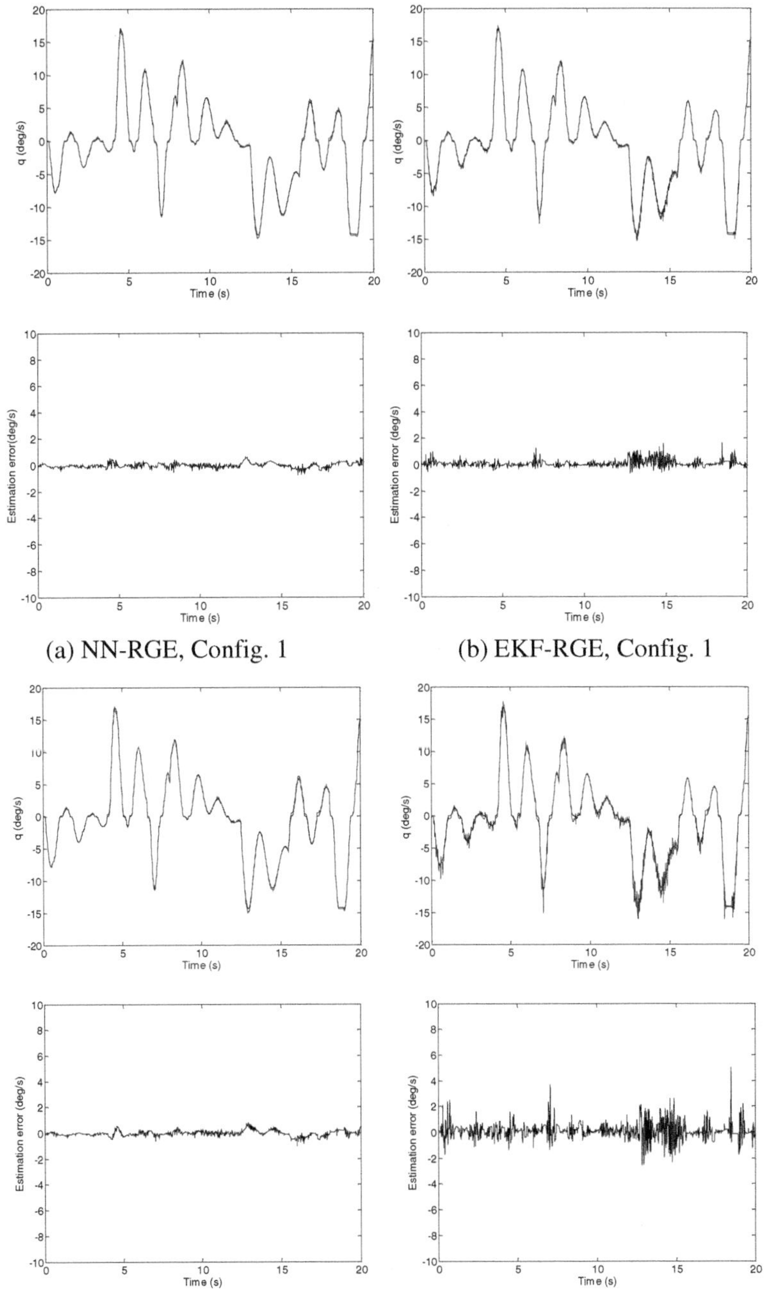

(a) NN-RGE, Config. 1 (b) EKF-RGE, Config. 1

(c) NN-RGE, Config. 6 (d) EKF-RGE, Config. 6

Fig. 5.14 NN and EKF estimations and estimation errors (basic difference, $\boldsymbol{q_{real}} - \boldsymbol{\hat{q}}$). Purple line is the NN and EKF estimates

Table 5.9 Fault Detection results summary

	Performance Indicator	Config 1	Config 2	Config 3	Config 4	Config 5	Config 6	Config 7
UD	NN-RGE	4	4	4	4	4	4	4
	NN-RGPE	0	0	0	0	0	1	1
	EKF-RGE	0	0	0	0	3	0	0
	EKF-RGPE	0	0	0	0	0	0	0
DR	NN-RGE	12.27	12.27	11.96	11.96	12.27	7.25	7.25
	NN-RGPE	40.23	40.23	31.32	31.32	40.23	23.40	23.40
	EKF-RGE	11.33	12.01	6.57	6.45	11.74	1.40	0.24
	EKF-RGPE	104.72	98.77	93.92	53.48	103.45	16.19	0.06
MT	NN-RGE	0.63	0.63	0.64	0.64	0.64	0.67	0.67
	NN-RGPE	1.29	1.29	1.35	1.35	1.29	1.54	1.21
	EKF-RGE	1.74	1.70	1.74	1.68	1.36	1.21	1.74
	EKF-RGPE	3.04	2.80	3.12	3.10	3.19	4.75	5.06
FA	NN-RGE	0	0	0	0	0	0.09	0.09
	NN-RGPE	0	0	0	0	0	0	0
	EKF-RGE	0.09	0.25	0.20	0.41	1.14	1.68	2.32
	EKF-RGPE	0	0	0	0	0	0.77	1.23

UD: Number of undetected faults
DR: Detectability ratio (see Eq. 5.20)
MT: Mean detection time in seconds
FA: Percentage false alarm (see Eq. 5.21)

Table 5.10 Fault detection time in seconds for Configuration 1

Fault type	NN-RGE	NN-RGPE	EKF-RGE	EKF-RGPE
CB_L	0.50	1.00	0.48	1.00
ST_L	0.50	1.00	0.50	1.00
HA_L	1.04	1.16	1.10	1.36
SA_L	-	1.42	2.68	3.10
CB_S	0.50	1.00	0.48	1.00
ST_S	-	1.00	2.86	5.62
HA_S	-	1.36	2.86	5.62
SA_S	-	2.42	3.00	5.62

-: Fault not detected

Table 5.11 Fault detection time in seconds for Configuration 2

Fault type	NN-RGE	NN-RGPE	EKF-RGE	EKF-RGPE
CB_L	0.50	1.00	0.48	1.00
ST_L	0.50	1.00	0.50	1.00
HA_L	1.04	1.16	1.06	1.36
SA_L	-	1.42	2.46	3.56
CB_S	0.50	1.00	0.48	1.00
ST_S	-	1.00	2.86	4.82
HA_S	-	1.36	2.86	4.82
SA_S	-	2.42	2.94	4.82

-: Fault not detected

Table 5.12 Fault detection time in seconds for Configuration 3

Fault type	NN-RGE	NN-RGPE	EKF-RGE	EKF-RGPE
CB_L	0.50	1.00	0.48	1.00
ST_L	0.50	1.00	0.50	1.00
HA_L	1.08	1.18	1.10	1.52
SA_L	-	1.52	2.72	3.58
CB_S	0.50	1.00	0.48	1.00
ST_S	-	1.00	2.82	5.62
HA_S	-	1.38	2.82	5.62
SA_S	-	2.74	2.96	5.62

-: Fault not detected

Table 5.13 Fault detection time in seconds for Configuration 4

Fault type	NN-RGE	NN-RGPE	EKF-RGE	EKF-RGPE
CB_L	0.50	1.00	0.48	1.00
ST_L	0.50	1.00	0.50	1.00
HA_L	1.08	1.18	1.08	1.36
SA_L	-	1.52	2.42	3.58
CB_S	0.50	1.00	0.48	1.00
ST_S	-	1.00	2.80	5.62
HA_S	-	1.38	2.80	5.62
SA_S	-	2.74	2.90	5.62

-: Fault not detected

Table 5.14 Fault detection time in seconds for Configuration 5

Fault type	NN-RGE	NN-RGPE	EKF-RGE	EKF-RGPE
CB_L	0.50	1.00	0.48	1.00
ST_L	0.50	1.00	0.50	1.00
HA_L	1.04	1.16	1.22	1.36
SA_L	-	1.42	4.14	4.28
CB_S	0.50	1.00	0.48	1.00
ST_S	-	1.00	-	5.62
HA_S	-	1.36	-	5.62
SA_S	-	2.42	-	5.62

-: Fault not detected

Table 5.15 Fault detection time in seconds for Configuration 6

Fault type	NN-RGE	NN-RGPE	EKF-RGE	EKF-RGPE
CB_L	0.50	1.00	0.48	1.00
ST_L	0.50	1.00	0.50	4.82
HA_L	1.18	1.20	1.12	4.82
SA_L	-	1.78	2.40	4.82
CB_S	0.50	1.00	0.48	1.00
ST_S	-	1.00	2.40	7.20
HA_S	-	1.52	2.40	7.20
SA_S	-	-	2.54	7.20

-: Fault not detected

Table 5.16 Fault detection time in seconds for Configuration 7

Fault type	NN-RGE	NN-RGPE	EKF-RGE	EKF-RGPE
CB_L	0.50	1.00	0.48	1.00
ST_L	0.50	1.00	0.50	5.62
HA_L	1.18	1.20	1.22	5.62
SA_L	-	1.78	2.78	5.62
CB_S	0.50	1.00	0.48	1.00
ST_S	-	1.00	2.84	7.20
HA_S	-	1.52	2.84	7.20
SA_S	-	-	2.80	7.20

-: Fault not detected

Table 5.17 Fault detectability ratio (*DR*) for Configuration 1

Fault type	NN-RGE	NN-RGPE	EKF-RGE	EKF-RGPE
CB_L	51.13	173.62	47.05	482.13
ST_L	1.92	16.93	2.16	76.13
HA_L	1.68	2.37	1.83	11.67
SA_L	-	2.69	0.88	6.32
CB_S	43.43	116.81	35.46	219.93
ST_S	-	4.64	0.99	13.85
HA_S	-	3.00	0.99	13.85
SA_S	-	1.81	1.31	13.85

-: Fault not detected

Table 5.18 Fault detectability ratio (*DR*) for Configuration 2

Fault type	NN-RGE	NN-RGPE	EKF-RGE	EKF-RGPE
CB_L	51.13	173.62	49.50	443.17
ST_L	1.92	16.93	2.35	73.84
HA_L	1.68	2.37	2.03	12.26
SA_L	-	2.69	0.85	41.32
CB_S	43.43	116.81	37.35	203.42
ST_S	-	4.64	1.18	5.37
HA_S	-	3.00	1.18	5.37
SA_S	-	1.81	1.67	5.37

-: Fault not detected

Table 5.19 Fault detectability ratio (*DR*) for Configuration 3

Fault type	NN-RGE	NN-RGPE	EKF-RGE	EKF-RGPE
CB_L	49.94	144.01	26.99	446.86
ST_L	1.77	11.67	1.20	62.15
HA_L	1.44	1.67	1.04	18.18
SA_L	-	1.78	0.50	13.28
CB_S	42.51	84.81	20.26	179.13
ST_S	-	3.42	0.74	10.58
HA_S	-	2.05	0.74	10.58
SA_S	-	1.18	1.10	10.58

-: Fault not detected

Table 5.20 Fault detectability ratio (*DR*) for Configuration 4

Fault type	NN-RGE	NN-RGPE	EKF-RGE	EKF-RGPE
CB_L	49.94	144.01	26.70	251.64
ST_L	1.77	11.67	1.24	37.64
HA_L	1.44	1.67	1.08	4.15
SA_L	-	1.78	0.48	12.08
CB_S	42.51	84.81	20.08	101.69
ST_S	-	3.42	0.76	6.88
HA_S	-	2.05	0.76	6.88
SA_S	-	1.18	0.52	6.88

-: Fault not detected

Table 5.21 Fault detectability ratio (*DR*) for Configuration 5

Fault type	NN-RGE	NN-RGPE	EKF-RGE	EKF-RGPE
CB_L	51.13	173.62	49.50	443.17
ST_L	1.92	16.93	2.35	73.84
HA_L	1.68	2.37	2.03	12.26
SA_L	-	2.69	2.72	54.66
CB_S	43.43	116.81	37.35	203.42
ST_S	-	4.64	-	13.42
HA_S	-	3.00	-	13.42
SA_S	-	1.81	-	13.42

-: Fault not detected

Table 5.22 Fault detectability ratio (*DR*) for Configuration 6

Fault type	NN-RGE	NN-RGPE	EKF-RGE	EKF-RGPE
CB_L	30.54	104.27	5.95	87.12
ST_L	0.85	11.02	0.09	1.62
HA_L	0.76	4.59	0.20	1.62
SA_L	-	1.59	0.13	1.62
CB_S	25.83	60.61	4.45	32.73
ST_S	-	2.50	0.12	1.60
HA_S	-	2.63	0.12	1.60
SA_S	-	-	0.12	1.60

-: Fault not detected

Table 5.23 Fault detectability ratio (*DR*) for Configuration 7

Fault type	NN-RGE	NN-RGPE	EKF-RGE	EKF-RGPE
CB_L	30.54	104.27	0.88	0.18
ST_L	0.85	11.02	0.01	0.02
HA_L	0.76	4.59	0.02	0.02
SA_L	-	1.59	0.05	0.02
CB_S	25.83	60.61	0.65	0.07
ST_S	-	2.50	0.04	0.06
HA_S		2.63	0.04	0.06
SA_S	-	-	0.02	0.06

-: Fault not detected

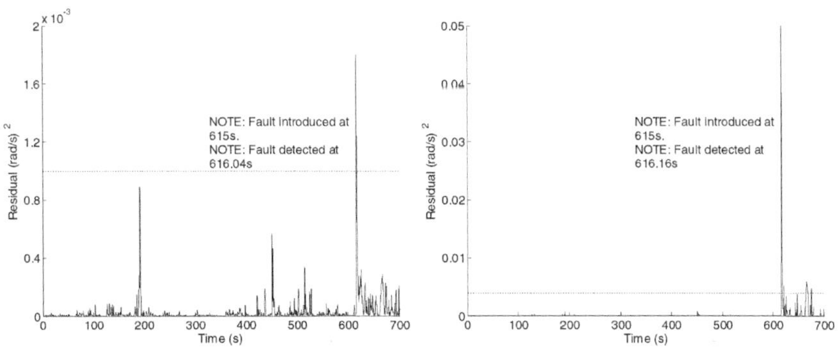

(a) NN-RGE, (HA_L) (b) NN-RGPE, (HA_L)

Fig. 5.15 Residual plots for Configuration 1. (RGE on the left, RGPE on the right)

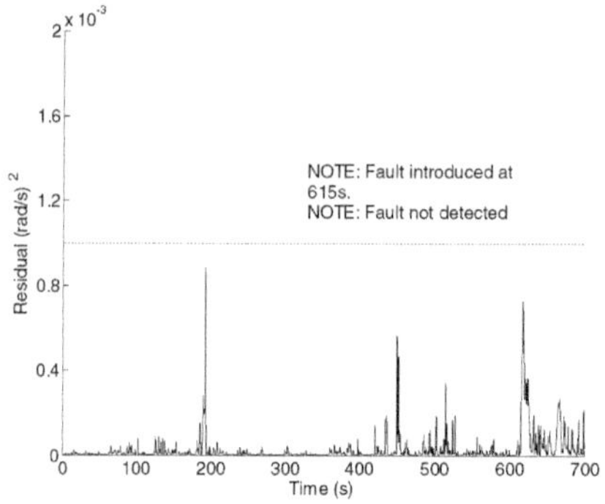

(c) NN-RGE, (SA_L)

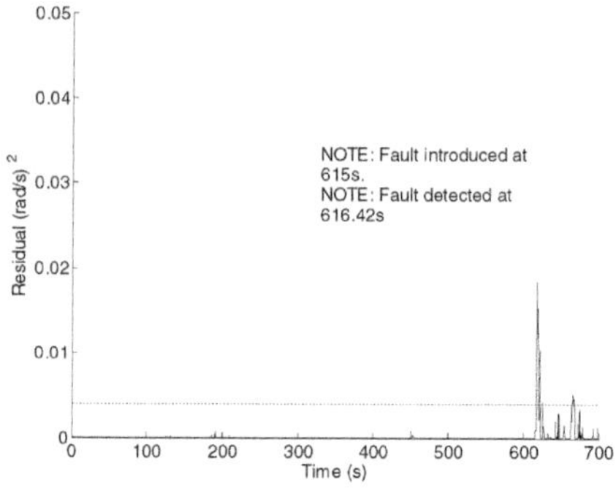

(d) NN-RGPE, (SA_L)

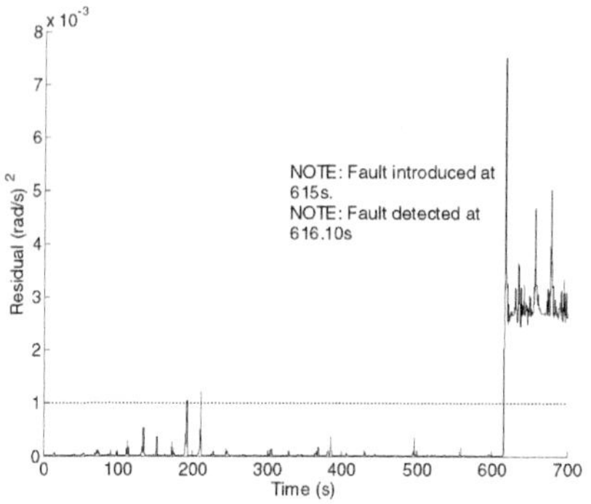

(e) EKF-RGE, (HA_L)

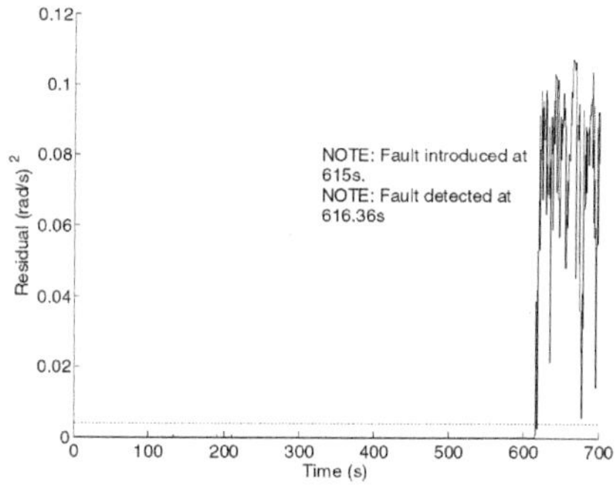

(f) EKF-RGPE, (HA_L)

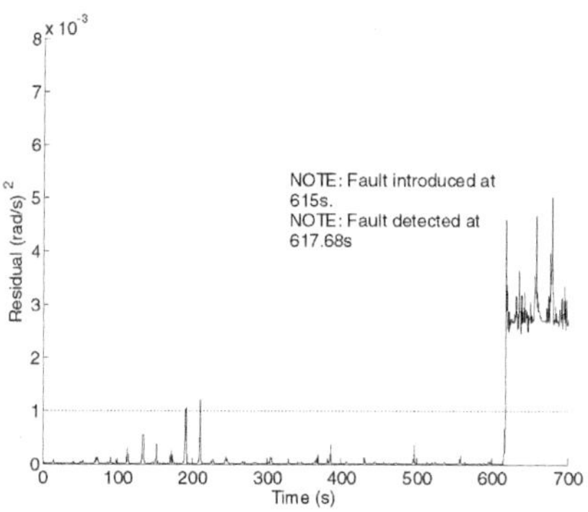

(g) EKF-RGE, (SA_L)

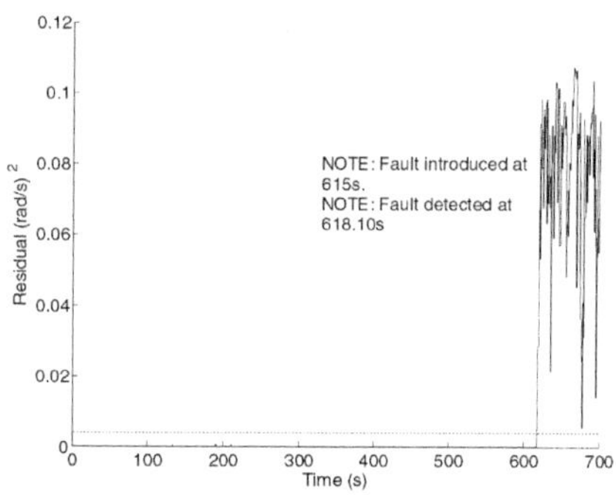

(h) EKF-RGPE, (SA_L)

Fig. 5.15 (*continued*)

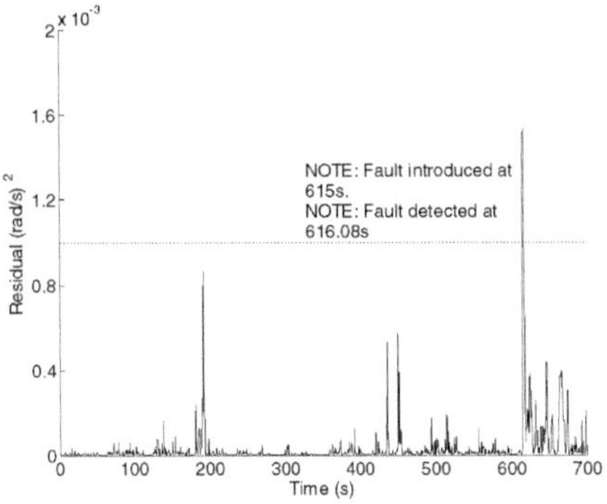

(a) NN-RGE, (HA_L)

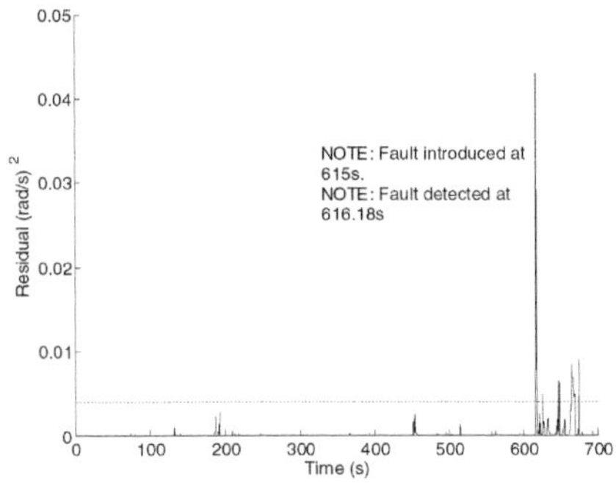

(b) NN-RGPE, (HA_L)

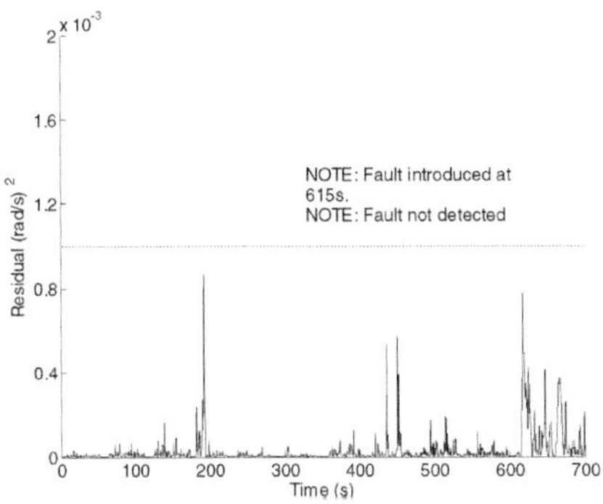

(c) NN-RGE, (SA_L)

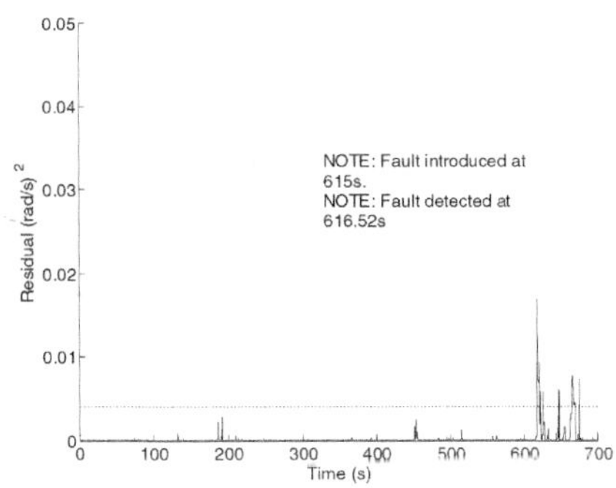

(d) NN-RGPE, (SA_L)

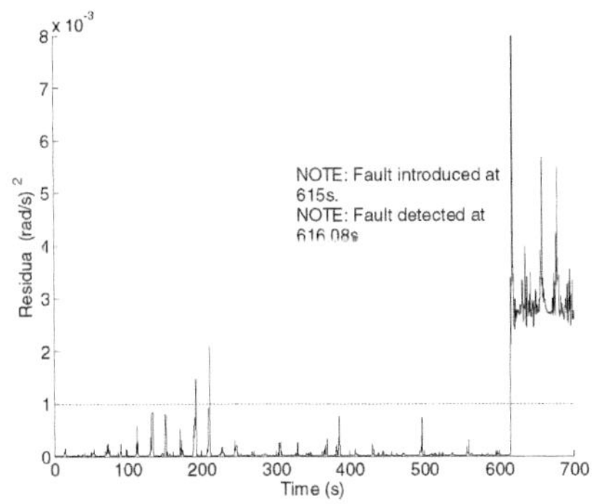

(e) EKF-RGE, (HA_L)

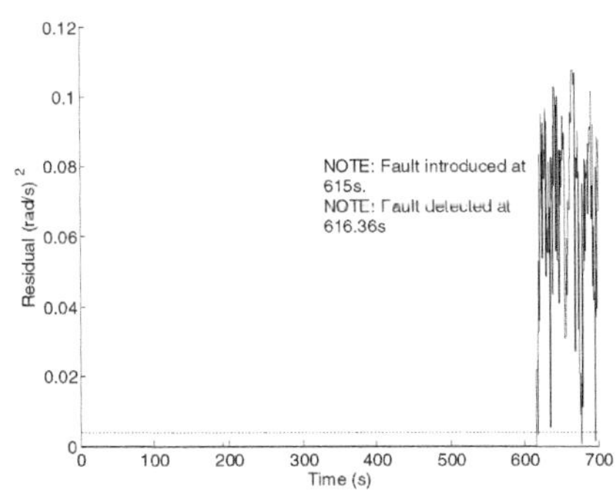

(f) EKF-RGPE, (HA_L)

Fig. 5.16 Residual plots for Configuration 4

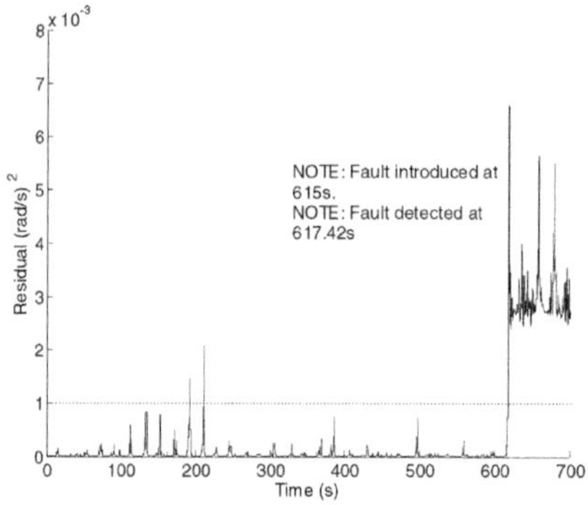

(g) EKF-RGE, (SA_L)

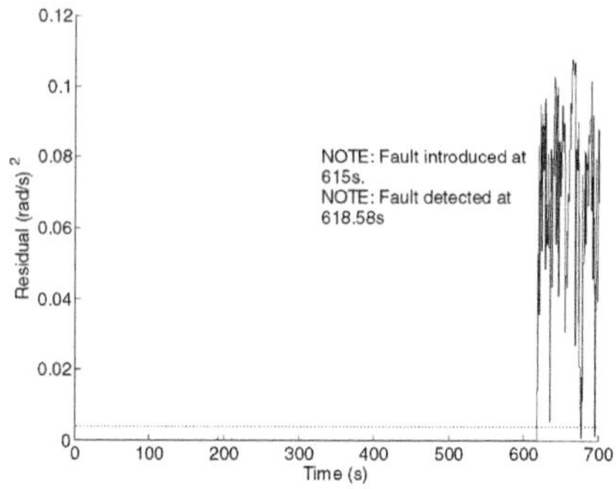

(h) EKF-RGPE, (SA_L)

Fig. 5.16 *(continued)*

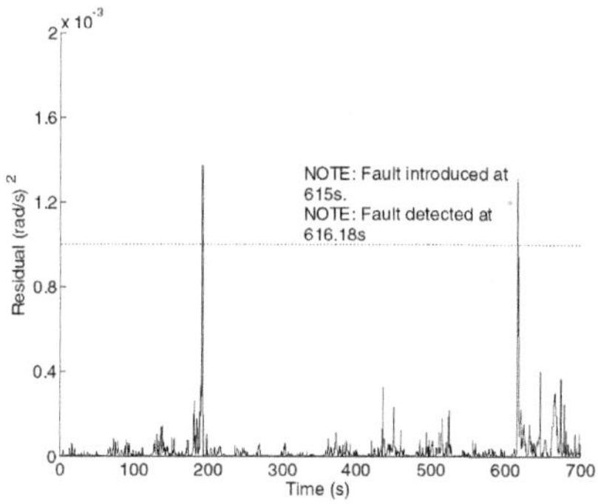

(a) NN-RGE, (HA_L)

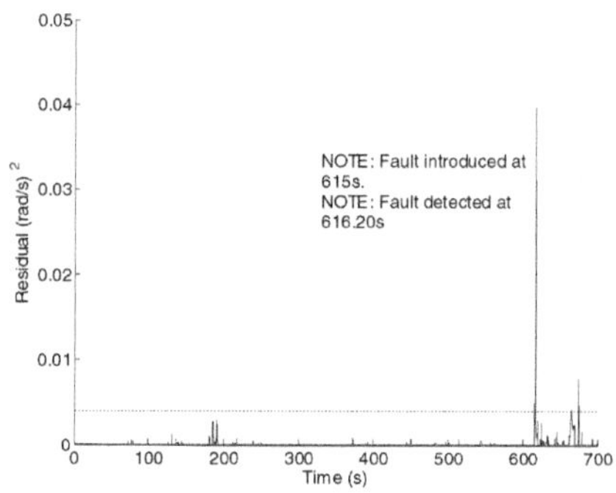

(b) NN-RGPE, (HA_L)

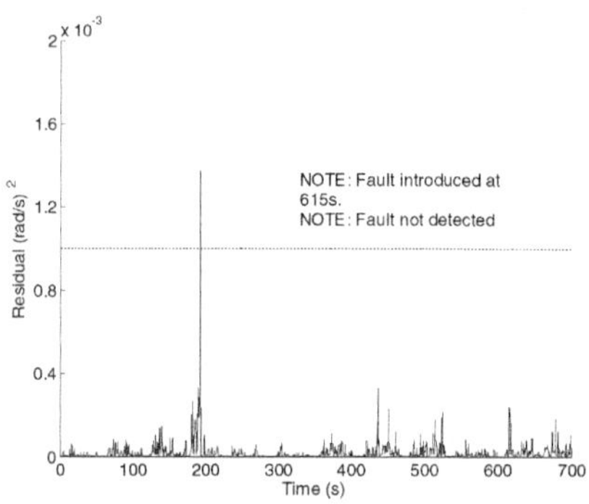

(c) NN-RGE, (SA_L)

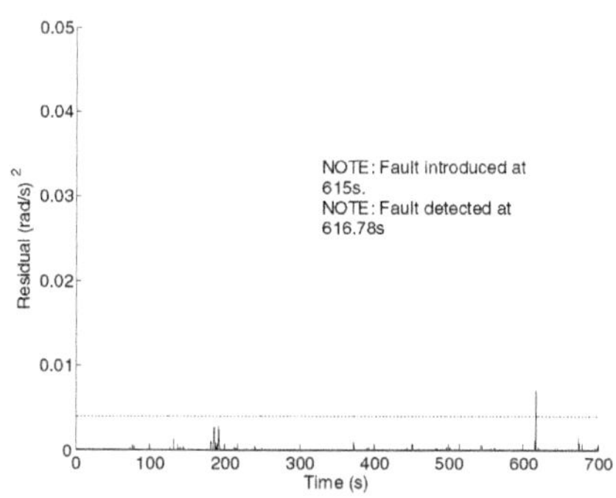

(d) NN-RGPE, (SA_L)

Fig. 5.17 Residual plots for Configuration 7

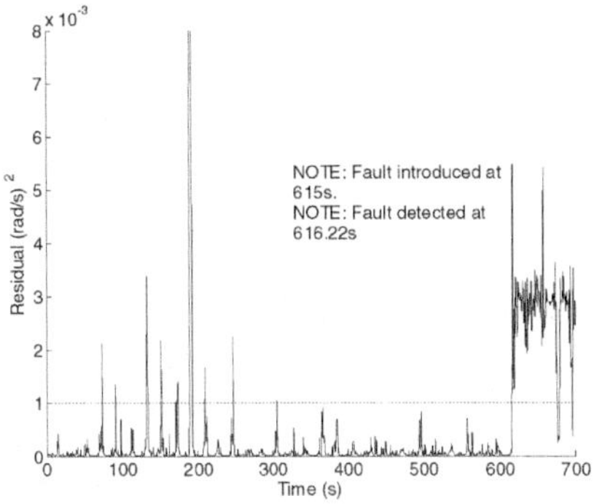

(e) EKF-RGE, (HA_L)

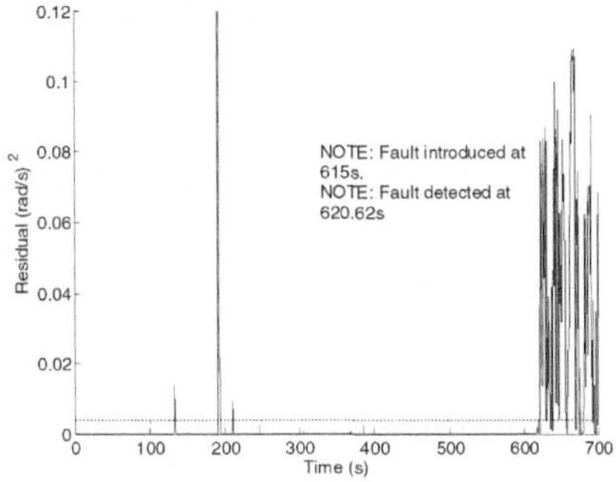

(f) EKF-RGPE, (HA_L)

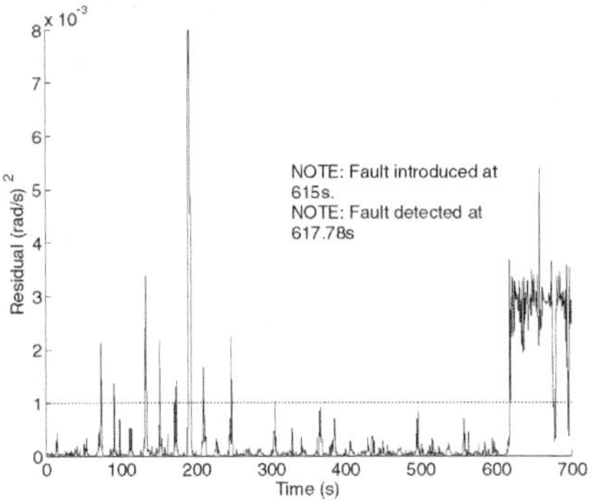

(g) EKF-RGE, (SA_L)

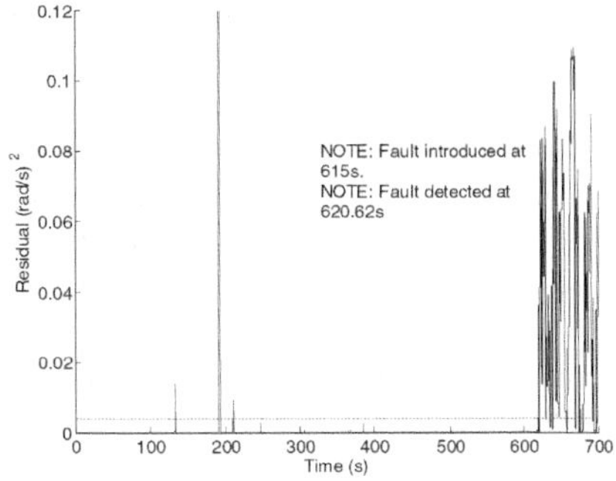

(h) EKF-RGPE, (SA_L)

Fig 5.17 (*continued*)

Table 5.24 Fault accommodation results summary. *MEE* and *VEE* are calculated from the time the fault is detected until the end of the data set.

	NN-based		EKF-based	
	MEE	*VEE*	*MEE*	*VEE*
Config. 1	0.39	0.12	0.27	0.20
Config. 2	0.39	0.12	0.30	0.23
Config. 3	0.42	0.12	0.36	0.36
Config. 4	0.42	0.12	0.38	0.39
Config. 5	0.39	0.12	0.49	1.98
Config. 6	0.42	0.11	0.77	1.84
Config. 7	0.42	0.11	0.83	3.28

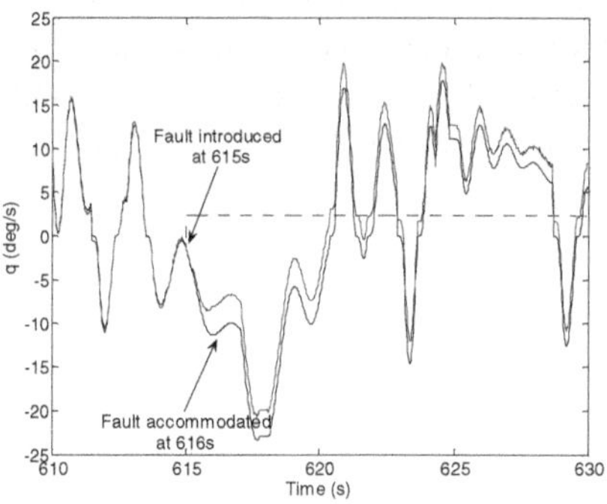

(a) NN-RGPE, (CB_L)

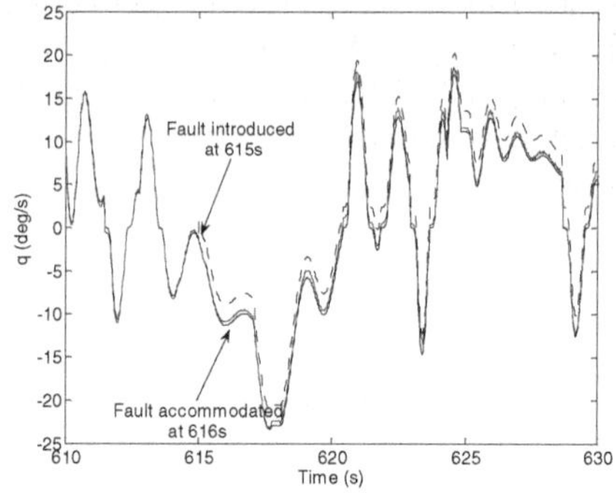

(b) NN-RGPE, (ST_L)

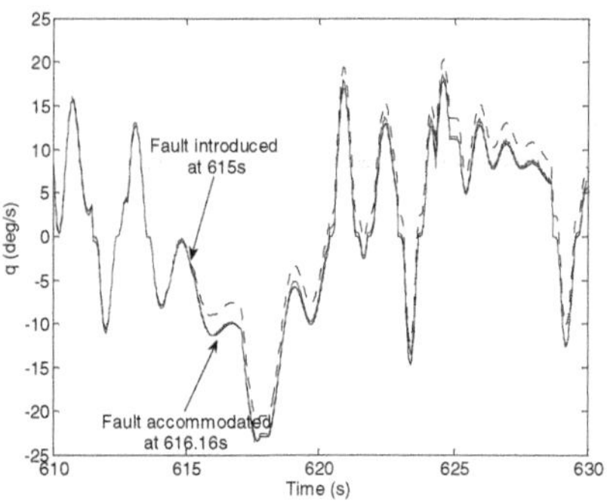

(c) NN-RGPE, (HA_L)

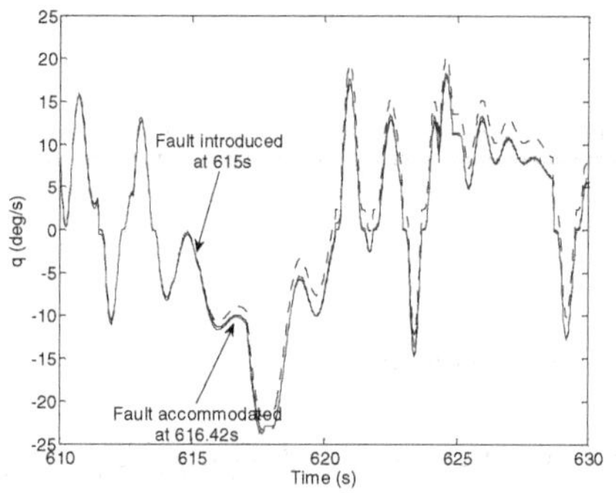

(d) NN-RGPE, (SA_L)

Fig. 5.18 NN fault accommodation plots for Configuration 1. All faults introduced at 615s; faulty (dotted), ideal (black), NN estimate (purple).

Conclusions

The aim of this chapter was to demonstrate a NN-based SFDA scheme on a UAV model, where only single sensor faults are considered. For comparative reasons, an EKF-based approach was also implemented. Tests were implemented on a nonlinear UAV model and faults were assumed to occur only in the pitch gyro. The tests were carried out under different levels of system and measurement noise and the EKF equations included parameter uncertainties. The following conclusions can be drawn:

- The EKF outperforms the NN, simply as a fault detector. By this we are implying that if we are only interested in detecting the faults and not the overall SFDA performance, then EKF-based schemes generally outperform NN-based schemes as the online training algorithm of NNs can potentially learn the faults. This was shown by the generally higher detectability ratios and the lower number of undetected faults for EKF-based SFDA schemes.

- On the other hand, NN-based SFDA schemes have lower false alarm rates than EKF-based SFDA schemes. This is because of the NN adaptive capabilities (via online training) which result in more accurate estimates.
- The NN has shown to be more robust to system and measurement noise than the EKF (provided that sufficient amount of offline training is implemented). This is due to their adaptive capabilities which allow them to adapt suitably to the time-varying system.
- A general guideline is that the average model estimation errors should be lower than the smallest expected sensor fault magnitude. This allows the residual generator to differentiate between a fault and a routine modelling error. In our tests only the EKF and the NN (but not the KF) were able to fulfil this criterion.
- NN-based SFDA schemes struggle to detect incipient faults as the NN inevitably learns the faulty measurements during training. Nevertheless with an appropriate residual generator, incipient faults can be successfully detected.
- In our case a weighted moving average (RGE) was used to compute the residual. Additionally a novel residual generator was proposed (RGPE) which was similar to the standard RGE approach but included a novel processing technique referred to as padding. Overall RGPE outperformed RGE with fewer false alarms and undetected faults as well as higher fault detectability ratios. This observation was true for all levels of unknown inputs, as well as the method of modelling (i.e. NN or EKF).
- The performance of NN-based SFDA schemes can be highly dependent on the fault detection time. A slow fault detection time can result in significant contamination of the NN structure (due to the faulty data). The contaminated NN can in turn degrade the fault accommodation accuracy.
- Constant bias faults are generally detected more quickly than incipient faults and result in higher detectability ratios. On the other hand constant bias faults can also contaminate the NN faster than incipient faults resulting in poorer fault accommodation performance. Methods which can be applied to avoid this include; tuning the NN, e.g. by lowering the learning rate so its global approximation capabilities are improved, tuning the residual generator, decreasing the fault detection time by lowering the threshold level (however we must also avoid increasing the false alarm rate).

Some of the flight conditions we consider in this chapter are not realistic of real flight. For example, the repeated 3-2-1-1 input demand is not realistic, as the pilot may consider other input patterns such as; ramp up/down and fixed elevator demands. However in this chapter the resulting flight data (e.g. pitch rate, normal acceleration etc.) are different for the NN training set and the NN testing set. This is arguably more important at this stage than testing realistic flight conditions, as we are primarily interested in the performance of the NN when it is queried with new data. However further work (as will be discussed later) can consider more realistic flight scenarios. The EKF was chosen here as a representative of SFDA schemes which are based on a mathematical description (fixed) of the system. In a number of occasions we have seen how the NN adaptive capabilities make them superior to the EKF. However it must be noted that satisfactory performance of an SFDA application depends on a number of contradicting properties and as such a trade-off will always be necessary. For example while the NN-based SFDA schemes have more accurate estimations which result in lower false alarms rates, the EKF-based SFDA schemes generally have fewer undetected faults. Furthermore a trade-off is generally required between the need to lower the residual threshold (to avoid undetected faults) and the need to raise the threshold (to avoid false alarms). In this chapter we have managed to tune an EMRAN RBF NN and design a novel residual generator (RGPE) so as to achieve on average; zero false alarm rate, no undetected faults, a 1.33s fault detection time, a fault accommodation error of 0.41 deg/s and an execution time of 0.69ms per sample of data (when implemented on a 1.6GHz Pentium processor). As a general remark, the conclusions drawn from this chapter are not intended to demonstrate the superiority of NNs to all mathematical-model based approaches. Rather the EKF was chosen due to its popularity and is used as a benchmark for which the NN performance can be compared to. In the next chapter, we use the NN and the residual structure designed here (NN-RGPE) to carry out a more realistic SFDA application where multiple sensor fault scenarios are considered.

Chapter 6
SFDIA-Multiple Sensor Faults

Introduction

A NN-based SFDIA scheme is proposed for the detection of multiple sensor faults in a nonlinear UAV model. In general, a group of sensors can fail simultaneously or consecutively (i.e. one at a time). In this chapter, we consider the latter scenario where a 1s time gap is allowed between consecutive sensor faults. However as before we assume that once a sensor has failed it cannot recover from the fault. With reference to Chapter 5, the NN-RGPE approach is implemented on the same nonlinear UAV model however sensor faults are now introduced in the pitch gyro, angle of attack sensor and normal accelerometer. The tests are carried out under different sensor fault types (step-type, hard and soft additive faults) but only using Configuration 1 settings (refer to Chapter 5, section 5.8.2).

In multiple sensor fault applications it is important that the faults are isolated (located) and not simply detected. This allows us to designate appropriate maintenance action such as replacing the faulty sensor with a redundant 'healthy' sensor. As such the overall scheme is more conveniently referred to as sensor fault detection, isolation and accommodation (SFDIA).

In general it is unlikely that only one sensor will fail in a real system. Therefore it is important that the SFDIA scheme is designed to detect multiple sensor faults. There are already several well-established SFDIA schemes in the literature, some of which have been discussed in Chapter 2. These include the dedicated observer scheme (DOS), generalised observer scheme (GOS) and the multiple model Kalman filter (MMKF). SFDIA schemes based on NNs have also been proposed particularly by the research group around Napolitano [118, 24-28]. In [143], the author suggests the use of a main NN (MNN) to detect a fault and a decentralised NN (DNN) to isolate the fault. In a design similar to the GOS, the *i*th-DNN is driven by all sensors except the *i*th sensor. Therefore if the *i*th sensor is faulty, the MNN and only the *i*th-DNN trigger a fault alarm. Other SFDIA applications using NNs include [19, 21, 23, 142]. Few however have implemented an EMRAN RBF NN to a UAV model. In Chapter 1 we emphasised that NN-based SFDIA schemes can be an invaluable alternative (in UAV applications), to physical redundancy due to space, weight and cost implications. Furthermore NN-based methods benefit from online learning capabilities, in comparison to traditional approaches such as the EKF (see Chapter 5), which can be useful in time-varying systems.

This chapter follows the same format as Chapter 5 (i.e. with the same subheadings and figures) for clarity purposes. As before, the results are grouped at the end of the chapter.

I. Samy and D.-W. Gu: Fault Detection and Flight Data Measurement, LNCIS 419, pp. 83–107.
springerlink.com

6.1 General SFDIA Outline and Terminologies

As in Chapter 5 (section 5.1), a sensor 'x' is referred to as Sensor-x, a sensor measurement is denoted by subscript '*real*', a NN estimate is denoted by a '$\hat{}$' and a subscript NN, and an ideal (non-faulty) sensor measurement is denoted by a subscript '*ideal*'.

SFDIA of multiple sensor fault scenarios can be divided into 2 stages. Consider Fig 6.1 where any of the *m* sensors can be faulty. Stage 1 involves modelling the different sensors where for example system model *m* includes all sensor measurements except Sensor–ym. This method of connection is in fact the well known GOS discussed in Chapter 2 except that a NN is used instead of an observer and the fault decision logic is different. Stage 2 is the residual processing stage which consists of a residual generator and simple threshold logic. SFDIA is implemented as follows (with reference to Fig 6.1):

- **Fault detection:** A fault in one sensor results in its corresponding residual to exceed the threshold. So for example a fault in Sensor-y1 will cause the first residual to trigger a fault alarm.
- **Fault isolation:** The sensor fault can be isolated by identifying which residual exceeded its threshold. For example if the first residual triggers a fault alarm then Sensor-y1 is faulty. The assumption behind this simple fault isolation logic is that if Sensor-y1 is faulty then the first residual will exceed the threshold **before** the faulty measurements contaminate system models 2, 3…m and cause the residuals 2, 3…m to also exceed their thresholds.
- **Fault accommodation:** Once a fault is detected and isolated, NN training is switched off and the faulty sensor is replaced with the NN estimate. So for example if Sensor-y1 is faulty, then system model 1 stops training and the first switch in Fig 6.1 shifts so that $y1_{real} = \hat{y}1$, where ideally $\hat{y}1 = y1_{ideal}$. The $\hat{y}1$ estimates are then used in the inputs of system models 2, 3…m.

There are three assumptions considered in our SFDIA tests. Firstly it is assumed that a sensor can only fail once, and this failure is permanent. Secondly, sensors can only fail one at a time with a 1s time gap between the consecutive faults. Finally for each test, it is assumed that all sensor fault types are the same. So for example if a step-type fault is introduced in Sensor-y1 then the next fault (e.g. in Sensor-y2) must also be step-type. This is assumed so that we can investigate the influence of each fault type separately.

In this chapter, sensor faults are assumed in Sensor-q, Sensor-α and Sensor- a_z. There are therefore 3 parallel NNs (system models) and 3 parallel residual processing units (i.e. consider Fig 6.1 with m = 3) which will be referred to as q-NN, α-NN, a_z-NN and q-RGPE, α-RGPE, a_z-RGPE respectively. Note that x-NN can refer to the SFDIA scheme performance as well as the NN model structure (see last two sentences in the statement below). To summarise the terminologies used in this chapter consider the following statement:

"If Sensor-q is faulty, then the difference between $\hat{q}_{NN}$ and q_{real} should cause q-RGPE to exceeds its threshold. q_{real} is then replaced with $\hat{q}_{NN}$ in the inputs of α -NN, a_z -NN and in the ideal scenario $\hat{q}_{NN} = q_{ideal}$. It was found that q-NN had a fault detection time of 1s. It was also found that q-NN consisted of 10 hidden neurons."

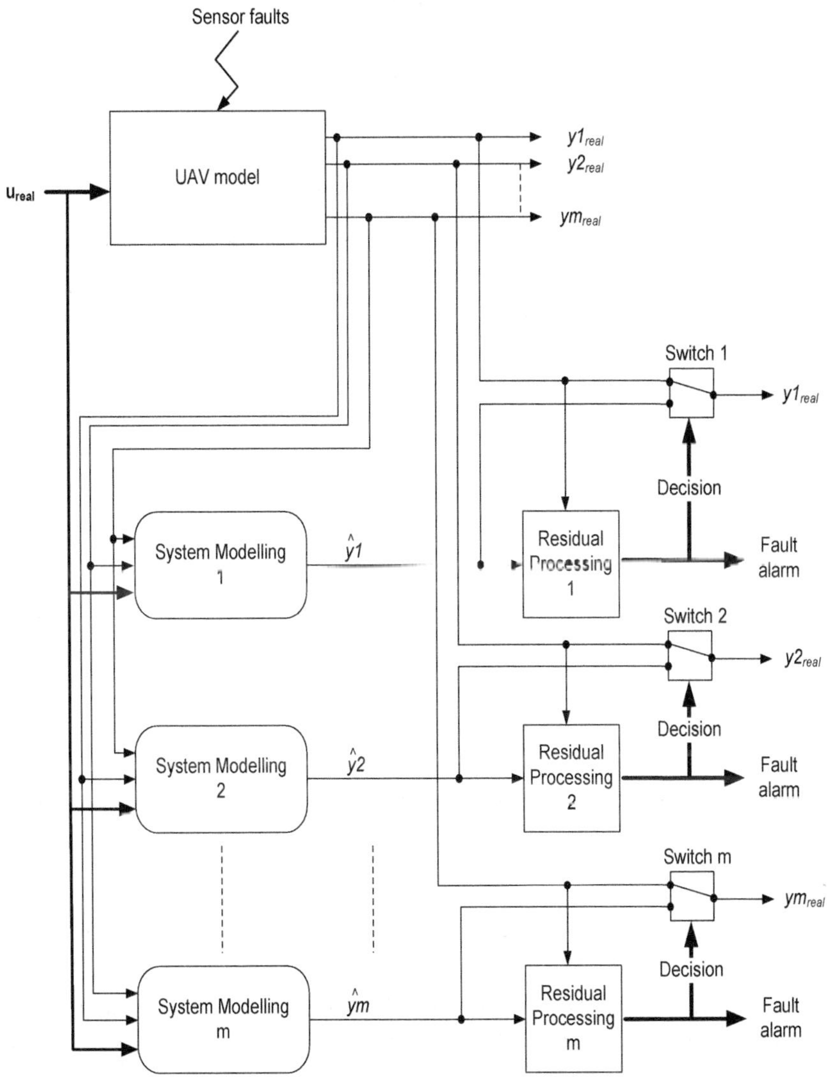

Fig. 6.1 General SFDIA outline for faults in Sensor-y1, Sensor-y2…Sensor-ym.

6.2 NNs Input/Output Structure

There are 3 NNs in the overall SFDIA scheme (section 6.1) which include q-NN, α-NN, a_z-NN and their outputs $\hat{q}_{NN}$, $\hat{\alpha}_{NN}$, $\hat{a}_{zNN}$ respectively. To maintain consistency, the same longitudinal flight variables defined in Chapter 5 (section 5.6) are used in the input sets of the NNs here. The resulting NN input/output structures are as follows (the structures are also displayed in Fig 6.2 and defined in Table 6.2):

$$\hat{q}_{NNk} = NN(\alpha_{k-1}, a_{zk-1}, \dot{w}_{k-1}, V_{tk-1}) \tag{6.1}$$

$$\hat{\alpha}_{NNk} = NN(q_{k-1}, a_{zk-1}, \dot{w}_{k-1}, V_{tk-1}) \tag{6.2}$$

$$\hat{a}_{zNNk} = NN(q_{k-1}, \alpha_{k-1}, \dot{w}_{k-1}, V_{tk-1}) \tag{6.3}$$

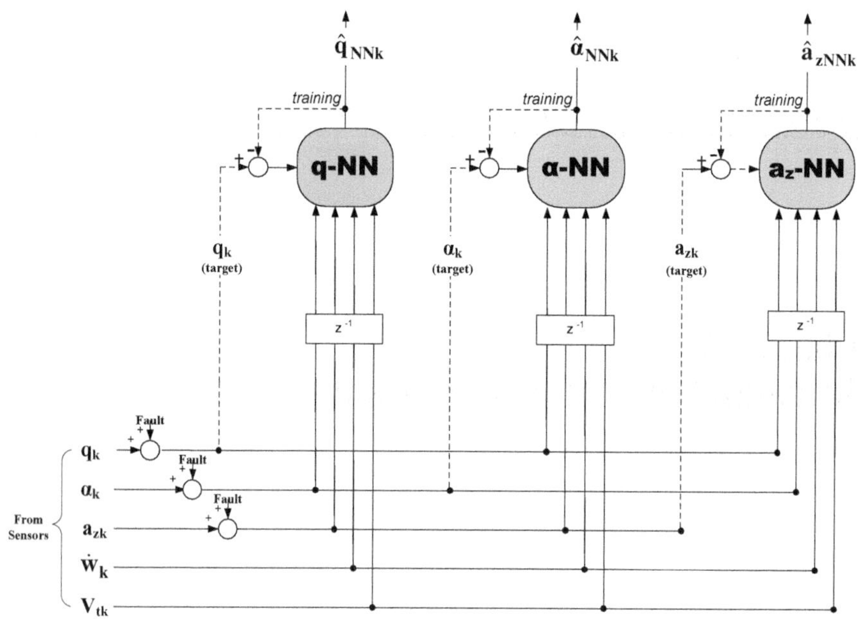

Fig. 6.2 q-NN, α-NN and a_z-NN input/output structures. Sensor faults can be present only in Sensor-q, Sensor- α and Sensor- a_z

6.3 Sensor Fault Types

Large additive faults are only considered in the SFDIA tests which include step-type ($T_R \cong 0$s), soft ($T_R = 4$s) and hard ($T_R = 1$s) additive faults. The magnitude of the fault (i.e. A) depends on which sensor is faulty. In Chapter 5 (section 5.7) a large fault in Sensor-q was assumed to have A=2.4 deg/s. For a ±30 deg/s pitch gyro sensor range this is 8% of the positive sensor range. Therefore to maintain consistency, an 8% sensor fault magnitude for Sensor-α and Sensor-a_z were considered resulting in A= 2.4 deg and A=1.25 m/s^2 respectively.

As discussed earlier (section 6.1) it is assumed that if the first sensor faulty type is e.g. step type then the second, third, etc. will also be step-type. Fig 6.3 shows an example of a $q \rightarrow \alpha$ fault sequence (i.e. Sensor-q faulty followed by Sensor-α faulty). Note how a 1s time gap is allowed between the time the first fault reaches its maximum value and the time the second fault is introduced. Unlike Chapter 5 where a fault is introduced at 615s, we now introduce the first fault at 626s to check the consistency of the fault detection scheme.

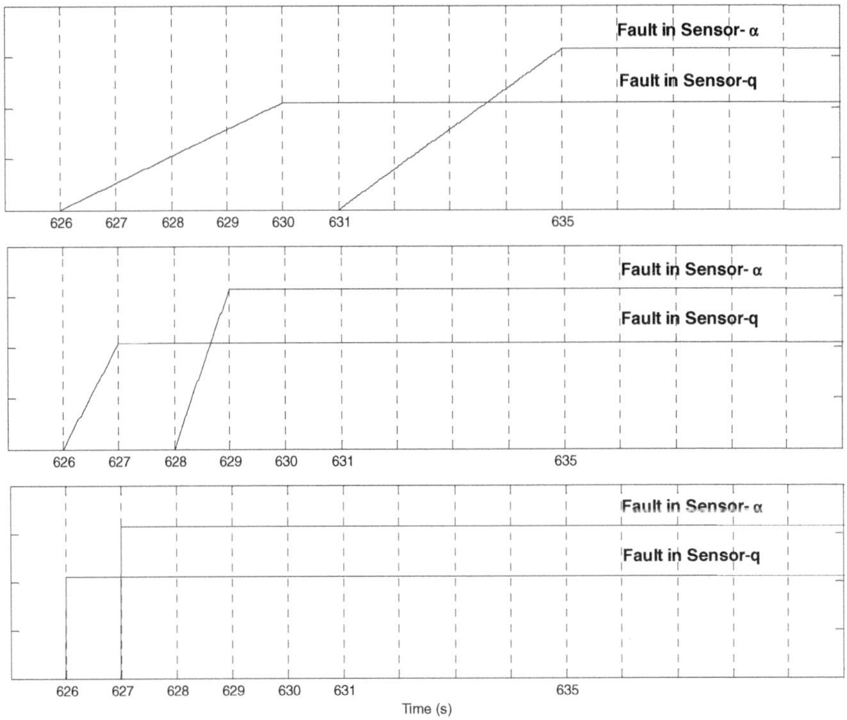

Fig. 6.3 Fault sequence $\boldsymbol{q} \rightarrow \boldsymbol{\alpha}$ (ignore fault magnitude). Top plot, middle plot and bottom plot are soft additive ($T_R = 4$s), hard additive ($T_R = 1$s) and step type ($T_R \cong 0$s) faults.

6.4 SFDIA Application to UAV Model

6.4.1 NN Training

The NN structure chosen in the SFDIA scheme is the EMRAN RBF NN discussed in Chapter 4. It is important that each of the NN models (q-NN, α-NN, a_z-NN) is initialised to avoid any poor initial estimations. Therefore each of the NN models is initialised separately via offline training based on the stopping criteria defined in Chapter 4 (section 4.2) and using 225s of UAV flight data. They are then used in the SFDIA scheme where their structures are continuously updated via online training.

6.4.2 SFDIA Test Outline

Flight data from Configuration 1 (Chapter 5, section 5.8.2) is used in the SFDIA tests and only 3 fault types are tested; large step-type, hard additive and soft additive faults (ST_L, HA_L, SA_L respectively). The following 10 fault sequences are implemented, where each sequence is considered a separate test:

- **Fault sequence 1:** *none*
- **Fault sequence 2:** $q \to none$
- **Fault sequence 3:** $\alpha \to none$
- **Fault sequence 4:** $a_z \to none$
- **Fault sequence 5:** $q \to \alpha$
- **Fault sequence 6:** $\alpha \to q$
- **Fault sequence 7:** $q \to a_z$
- **Fault sequence 8:** $a_z \to q$
- **Fault sequence 9:** $\alpha \to a_z$
- **Fault sequence 10:** $a_z \to \alpha$

where for example $q \to none$ implies that only Sensor-q is faulty, while $q \to \alpha$ implies that Sensor-q **followed** by Sensor-α are faulty. Fig 6.3 shows an example of the $q \to \alpha$ fault sequence.

In conclusion there are 30 separate tests (1 configuration x 3 fault types x 10 fault sequences x 1 SFDIA scheme) and the NN-RGPE structure (defined in Chapter 5) is used in the SFDIA scheme but with different tuned parameters than in Chapter 5 (Table 6.2). Fig 6.4 shows a tree-diagram for the different tests where each full branch represents one test. So for example one of the branches is the test where Sensor-q followed by Sensor-α ($q \to \alpha$ sequence) are faulty and both fault types are HA_L.

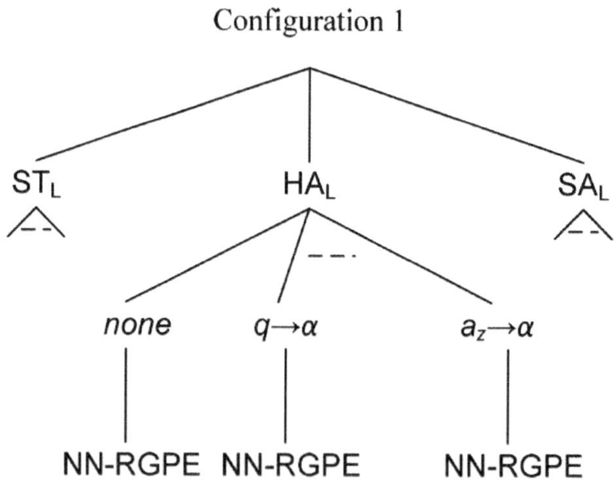

Fig. 6.4 Tree-diagram showing the different SFDIA tests. 1 configuration, 3 fault types, 10 fault sequences and 1 SFDIA scheme. Overall there are 30 separate tests.

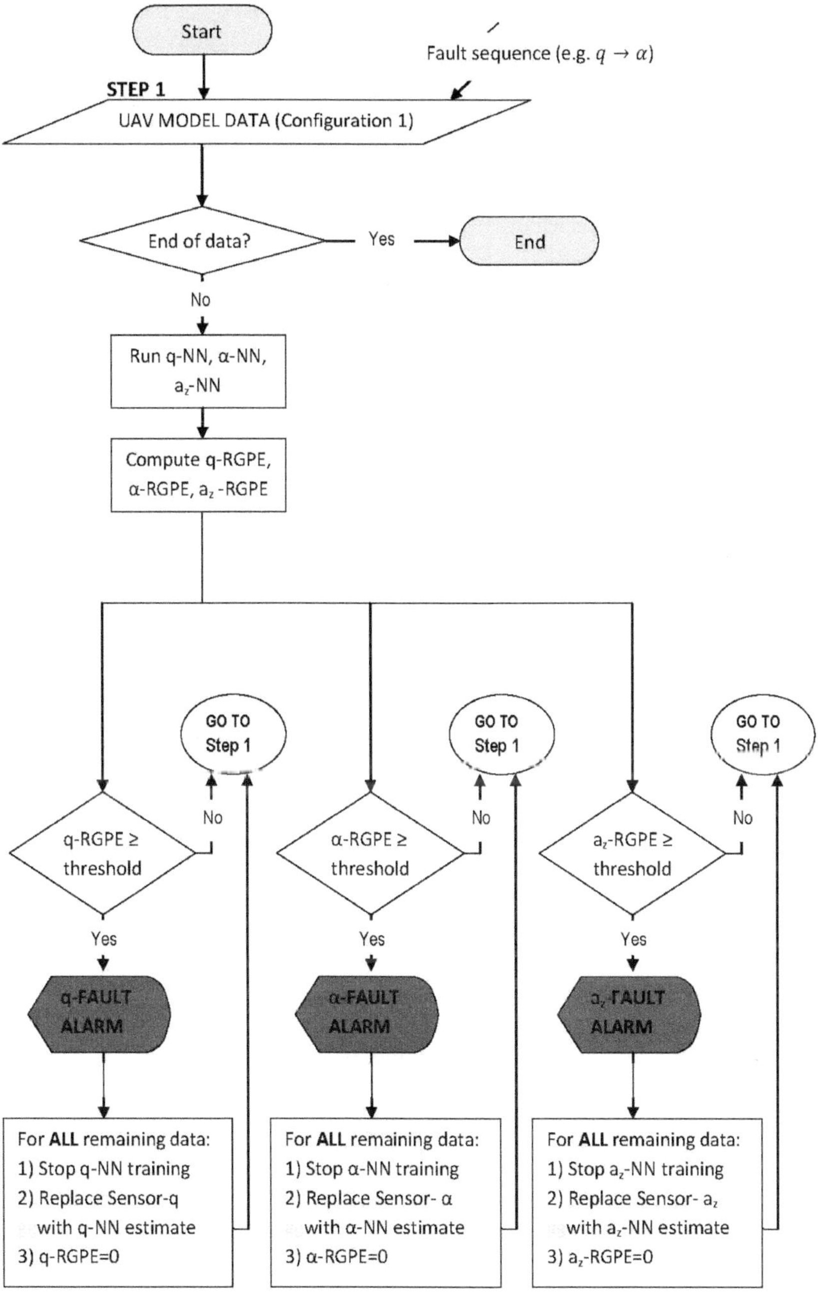

Fig. 6.5 Flow chart for SFDIA tests (one pass through the flowchart per sample instant)

Fig 6.5 shows the flowchart for the SFDIA tests where 1 pass through the flow chart (i.e. from Start to End) represents one test. Note that as we are assuming that each sensor can only fail once, the fault alarm is triggered only once for each sensor, which is why we set RGPE to zero once a fault is detected (see last section of the flowchart). This is important if we are to isolate the fault according to the fault isolation logic in Table 6.1 (i.e. so that consecutive faults are detected).

Table 6.1 Fault isolation logic

	Sensor-q failure	Sensor-α failure	Sensor-a_z failure
q-RGPE exceeds?	YES	NO	NO
α-RGPE exceeds?	NO	YES	NO
a_z-RGPE exceeds?	NO	NO	YES

6.4.3 SFDIA Performance Indicators

The *MEE* and *VEE* used in Chapter 5 (section 5.8.3) are also used here to assess the NN estimation performance. There are 3 possible outcomes from each SFDIA tests. Outcome 1 is when all faults are detected and isolated successfully; in this case we refer to the result as FD. Outcome 2 is when all faults are detected and isolated successfully but there is also a false alarm in one (or more) of the other residuals; in this case we refer to the result as FD-FA. Outcome 3 is when one (or more) of the faults are not detected; in this case we refer to the result as FND. These terms are only used in Fig 6.7 where FD is the ideal outcome and FND is the worst case scenario.

6.5 Results

The three NN models and RGPE structures used in the SFDIA scheme are outlined in Table 6.2. Together Tables 6.2-6.7 and Figs 6.6-6.14 display the SFDIA results:

- **Table 6.3:** This shows the number of training epochs required before the offline training stopping criteria (Chapter 4, section 4.2) are satisfied. A higher learning rate was used to train α-NN which could be the reason for α-NN requiring the least number of training epochs (only 1000 epochs). Sensor-α includes both system (gust disturbances) and measurement noise (Chapter 5, section 5.3.3) in comparison to other sensors which only include measurement noise. Therefore, α-NN must be designed with a higher learning due to the highly, noise-corrupted Sensor- α.
- **Fig 6.6:** As in Fig 5.13 (Chapter 5) this shows the mean execution times of each NN. In general we find that q-NN, α-NN and a_z-NN have almost the same execution times. More importantly the maximum execution time of the 3 NNs (working in parallel) must be less than the flight data sampling time if we are to avoid any time delays. α-NN had the highest processing time per data sample (0.55 ms) which is lower than the flight data sampling time (0.02s).

- **Table 6.4 and Fig 6.7:** These show the FDI results. In Table 6.4 the fault detection times are shown and a false alarm is denoted by 'False'. For example in the step-type $q \rightarrow \alpha$ sequence (i.e. where Sensor-q is first faulty followed by Sensor-α) the fault in Sensor-q is not detected, the Sensor-α fault is detected (in 1.00s) and a false alarm is present in the a_z-NN branch (i.e. a_z-RGPE exceeds its threshold). Fig 6.7 displays the same results, but graphically (to make it easier to visualise the FDI patterns). Three outcomes are possible in Fig 6.7 as discussed in section 6.4.3; FD (fault detected), FD-FA (fault detected but false alarm also present) and FND (fault not detected). Each cell in Fig 6.7 displays the FDI pattern (i.e. the order at which the residuals, q-RGPE, α-RGPE, a_z-RGPE, exceed their thresholds) for one fault scenario. For example in the soft-additive $q \rightarrow a_z$ sequence, q-RGPE exceeds its threshold **followed** by a_z-RGPE. Therefore in this case the sensor faults are both detected (FD outcome). On the other hand consider the step-type $\alpha \rightarrow q$ sequence. In this case α-RGPE first exceeds its threshold, followed by a_z-RGPE, followed by q-RGPE. Therefore the faults in Sensor- α and Sensor-q are detected however a_z -RGPE also shows a false alarm (FD-FA outcome). The ideal outcome is FD, and the worst case scenario is FND. In general we can see that step-type faults result in a larger number of false alarms and undetected faults (78% of the step-type tests resulted in FD-FA or FND outcomes) in comparison to soft and hard additive faults.
- **Table 6.5:** This shows the fault detection results for a three fault sequence $a_z \rightarrow \alpha \rightarrow q$. Two fault magnitudes are considered in Sensor-q; 2.4 deg/s or a large 15 deg/s. First we notice that faults in Sensor- a_z and Sensor-α are always detected. However only the 15 deg/s fault in Sensor-q is detected. It was found that the accommodated measurement from a_z-NN (i.e. in the event when a_z-NN estimates replace Sensor- a_z) had a large influence on the fault detection performance of q-NN. In fact if the a_z-NN accommodated measurements were replaced with nominal (fault-free) measurements then the 2.4 deg/s faults in Sensor-q are detected. In conclusion the performance of q-NN greatly depends on the performance of a_z-NN and vice versa.
- **Table 6.6:** This shows the fault accommodation results. The SFDIA tests are run and at the end of each test, the difference between the NN estimate (e.g. $\hat{q}_{NN}$) and ideal (e.g. q_{ideal}) measurement is calculated. For example in Table 6.6, the q-NN estimations are poor for the hard additive, $\alpha \rightarrow a_z$ fault sequence (estimation error of 7.12 deg/s).
- **Table 6.7:** This summarises the fault detection and accommodation results. For example q-NN on average detected soft additive faults in 1.82s. In general step-type faults result in faster fault detection times but poorer fault accommodation performance in comparison to soft additive and hard additive faults.
- **Fig 6.8-6.14:** These show the residual (q-RGPE, α-RGPE and a_z-RGPE) and NN estimation plots for the $q \rightarrow \alpha$ and $a_z \rightarrow \alpha \rightarrow q$ fault sequences. Ideally the NN estimates (purple lines) should closely follow the fault-free measurements (black lines). The dotted lines are the faulty measurements. In Fig 6.8 we find that the residuals do not exceed their thresholds when faults are not present. This indicates that 1) The NN estimations are accurate 2) The SFDIA scheme is robust to system or measurement noise (otherwise false alarms would be present).

Discussion

To avoid confusion let us first re-iterate the SFDIA sequence of operation when a fault is present in one of the sensors. Consider the SFDIA scheme in Fig 6.2. Let us assume that a fault in Sensor-q is introduced. Therefore the following would be the sequence of operation in the *ideal* scenario:

1. The fault in Sensor-q is seen in the feedback (training) connection of q-NN. Therefore q-NN would react to the fault by suddenly outputting inaccurate estimations. These inaccurate estimations along with the faulty sensor measurements will cause q-RGPE to exceed its threshold and a fault alarm to be triggered. Therefore the fault in Sensor-q is successfully detected.
2. Similarly the fault in Sensor-q is seen in the input set of α-NN and a_z-NN. However as the fault is only seen in the input set (and not the feedback connection as in q-NN), the performance of α-NN and a_z-NN may not be significantly degraded. Therefore the residuals α-RGPE and a_z-RGPE will not exceed the thresholds as desired, i.e. no false alarms.
3. So far q-RGPE has triggered a fault alarm. The next step would involve fault accommodation. In this case q-NN would stop training, and its estimates ($\hat{q}_{NN}$) would replace Sensor-q. Note that $\hat{q}_{NN}$ must be sufficiently accurate as it will be used (in place of Sensor-q) in the input set of α-NN and a_z-NN.

The last step shows how the NN-based SFDIA scheme is interconnected, i.e. the performance of one NN greatly affects the performance of other NNs especially during the fault accommodation stage.

In general step-type and constant bias faults are more quickly detected than drift-type faults. This is expected as the former have a more abrupt effect on the residual in comparison to drifting faults. This observation was clearly seen in Chapter 5 (see for example Tables 5.10 and 5.11) where step-type and constant bias faults were detected faster and with a higher detectability ratio than hard and soft additive faults. Again, this was also seen here (see e.g. Table 6.7) where step-type faults had faster fault detection times. However a side-effect which could not be seen in Chapter 5 was that step-type faults can result in poor fault detection and accommodation performances of the *overall* SFDIA scheme, i.e. a step-type fault in Sensor-q greatly affects the performance of α-NN and a_z-NN and not only q-NN. To explain this further, note the following two possible outcomes (referring to the SFDIA scheme, Fig 6.2):

Temporary NN contamination:

1. Sensor-q is faulty
2. q-RGPE has **not** yet exceeded its threshold
3. In the mean time α-NN and a_z-NN are gradually being contaminated from the faulty Sensor-q
4. q-RGPE exceeds its threshold and a fault alarm is triggered
5. q-NN stops training and its estimates ($\hat{q}_{NN}$) replace Sensor-q
6. As a result α-NN and a_z-NN are now receiving non-faulty measurements (i.e. $\hat{q}_{NN}$) and can recover from their *temporarily contaminated* structure.

Permanent NN contamination:

1. Sensor-q is faulty
2. q-RGPE has **not** yet exceeded its threshold
3. In the mean time α-NN and a_z-NN are gradually being contaminated from the faulty Sensor-q
4. This contamination causes α-RGPE to exceed its threshold resulting in a false alarm.
5. Because α-RGPE has exceeded its threshold, α-NN stops training and its structure is frozen. Therefore α-NN has become *permanently contaminated.*

Again, from the above we can see how the NN-based SFDIA scheme is highly interconnected and the ideal scenario would be to detect the faults as quickly as possible before they result in permanent NN contaminations. Therefore we can expect step-type faults to cause more damage to the SFDIA scheme due to their abrupt effect in comparison to incipient faults. This was repeatedly seen in our tests where 1) Fig 6.7 showed that step-type faults result in more false alarms and undetected faults 2) Table 6.7 showed that step-type faults result in larger fault accommodation errors, in comparison to soft and hard additive faults.

From the results we noted earlier that the performance of a_z -NN greatly depends on the performance of q-NN and vice versa. This dependency is best explained from the aircraft equations of motion. Normal acceleration is calculated as $a_z = \dot{w} - qu$ [140]. Therefore we can see that normal acceleration is dependent on the pitch rate and vice versa. This could be why a_z-NN is sensitive to faults in Sensor-q and similarly q-NN is sensitive to faults in Sensor-a_z.

There are several design solutions which we can adopt in order to reduce this sensitivity as well as improve the overall SFDIA scheme performance (e.g. so that step-type faults do not result in NN permanent contamination):

1. We can re-tune the residual structures RGPE so that faults are detected quicker.
2. We can increase the memory storage in the NN input (e.g.[q_{k-1} q_{k-2} .... q_{k-n}] instead of [q_{k-1}]) so that the NN is less sensitive to abrupt deviations in its inputs. However we must avoid increasing the NN processing time.
3. We can increase the number of NN input parameters so that the NN is less sensitive to one specific input parameter. However we must avoid increasing the NN processing time.
4. We can change the normalisation ranges of each parameter used in the NN input set. For example if the normalisation range for parameter 'q' (used in the input set of a_z –NN), is greatly increased then a_z -NN will become less sensitive to parameter 'q'. As a result of this sensitivity reduction, a_z -NN may not degrade in performance (i.e. no significant contamination) if Sensor-q is faulty.

The latter solution was in fact implemented in a separate test and it was found that if the normalisation limit (which was originally set to 0-1 in Table 6.2) is set at 0-10 (**only** for parameter 'q' used in the a_z-NN input set), then no false alarms were present in a_z-NN when faults were introduced in Sensor-q.

In general it was found that in addition to the SFDIA scheme structure, the performance of the SFDIA scheme greatly depends on the fault sign (i.e. positive or negative) and the time at which the fault is introduced. Due to insufficient time, the tests carried out in Chapter 5 and 6 only consider positive fault signs and faults were introduced randomly between 600-700s. However it is important that the SFDIA scheme is rigorously tested before implementation in a real system.

Table 6.2 Summary of NN and RGPE structures

		NN STRUCTURE	
	q-NN	α-NN	a_z-NN
Input parameters:	$[\alpha\ a_z\ \dot{w}\ V_t]$	$[\alpha\ a_z\ \dot{w}\ V_t]$	$[\alpha\ a_z\ \dot{w}\ V_t]$
Output parameter:	$\hat{q}$	$\hat{\alpha}$	$\hat{a}_z$
No. of input neurons:	0	0	0
No. of hidden neurons (max):	10	20	17
No. of output neurons:	1	1	1
Input data normalisation:	0-1	0-1	0-1
NN learning rate:	0.00005	0.00006	0.000001
$[E1\ E2\ \varepsilon_{max}\ \varepsilon_{min}\ \gamma_{dy}\ k_{op}]^a$:	[1e-2 0.01 0.6 0.3 0.997 1e-6]	[1e-3 0.01 0.6 0.3 0.997 1e-6]	[1e-3 0.01 0.6 0.3 0.997 1e-6]
		RESIDUAL STRUCTURE	
	q-RGPE	α-RGPE	a_z-RGPE
Weight $(\varpi)^b$:	91	200	15
Averaging size $(\Omega)^b$:	1	1	1
Padding points $(p_{pad})^b$:	1	1	1
Threshold:	0.004 $(rad/s)^2$	0.02 $(rad)^2$	0.2 $(m/s^2)^2$

a: Refer to Chapter 4, Eq. (4.11)-(4.16)
b: Refer to Chapter 5, Eq. (5.10)-(5.11)

Table 6.3 Offline training errors

	q-NN	α-NN	a_z-NN
Training epochs	2411	1000	3209
Estimation error	0.26 deg/s	0.39 deg	0.29 m/s^2

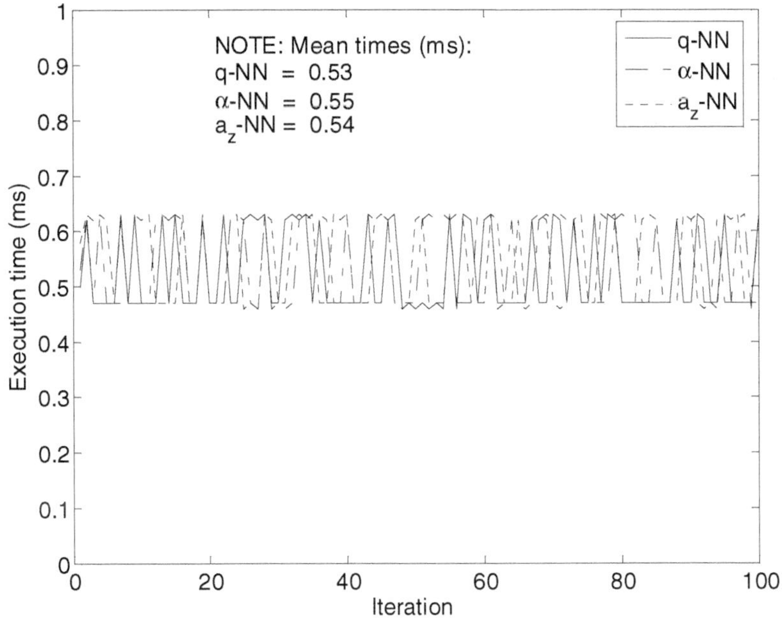

Fig. 6.6 Mean execution times per data sample. Mean calculated for first 100 data samples and then repeated for 100 iterations.

Table 6.4 Fault detection time (seconds) for the different fault sequences and faulty types. E.g. for $\boldsymbol{q} \rightarrow \boldsymbol{\alpha}$ sequence, soft additive fault type, the q-NN detects Sensor-q fault in 1.86s and α-NN detects Sensor-α fault in 2.86s. 'False' denotes a false alarm.

Fault Seq.	Soft additive			Hard additive			Step-type		
	q-NN	α-NN	a_z-NN	q-NN	α-NN	a_z-NN	q-NN	α-NN	a_z-NN
none	-	-	-	-	-	-	-	-	-
$q \rightarrow none$	1.86	-	-	1.24	-	-	1.00	-	False
$\alpha \rightarrow none$	-	1.86	-	-	1.16	-	-	1.00	-
$a_z \rightarrow none$	-	-	3.18	-	-	1.52	False	-	0.96
$q \rightarrow \alpha$	1.86	2.86	-	1.24	1.68	-	-	1.00	False
$\alpha \rightarrow q$	1.70	1.86	False	1.40	1.16	False	0.98	1.00	False`
$q \rightarrow a_z$	1.86	-	2.74	1.24	-	3.02	1.00	-	False
$a_z \rightarrow q$	-	-	3.18	0.66	-	1.52	False	-	0.96
$\alpha \rightarrow a_z$	False	1.86	1.64	False	1.16	1.40	False	1.00	0.32
$a_z \rightarrow \alpha$	-	2.12	3.18	-	1.22	1.52	-	1.00	0.96

-: Fault not detected

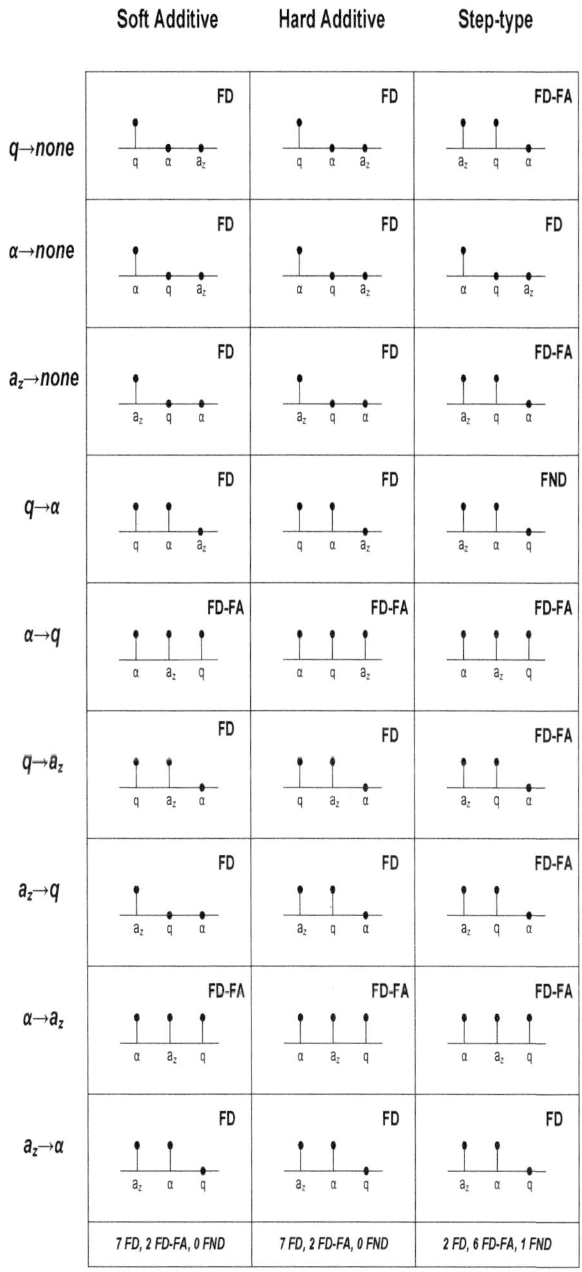

Fig. 6.7 Fault detection results for the different fault sequences and fault types. The horizontal line in each cell indicates which residual (q-RGPE,_α-RGPE, a_z-RGPE) exceeded its threshold and in what order. FD (Fault detected), FD-FA (Fault detected but False alarm present in other residuals), FND (Fault not detected).

Table 6.5 Fault detection time (seconds) for $\boldsymbol{a_z \rightarrow \alpha \rightarrow q}$ fault sequence

Fault Seq.	Soft additive			Hard additive			Step-type		
	q-NN	α-NN	a_z-NN	q-NN	α-NN	a_z-NN	q-NN	α-NN	a_z-NN
$a_z \rightarrow \alpha \rightarrow q$ [a]	-	2.12	3.18	-	1.22	1.52	-	1.00	0.96
$a_z \rightarrow \alpha \rightarrow q$ [b]	4.18	2.12	3.18	3.90	1.22	1.52	1.00	1.00	0.96

a: 2.4 deg/s fault magnitude for Sensor-q
b: 15 deg/s fault magnitude for Sensor-q

Table 6.6 Fault accommodation results. Estimation errors calculated for data between 626-635s and shown in deg/s, deg and m/s^2 for q-NN, α-NN and a_z -NN respectively.

Fault Seq.	Soft additive			Hard additive			Step-type		
	q-NN	α-NN	a_z-NN	q-NN	α-NN	a_z-NN	q-NN	α-NN	a_z-NN
none	0.16	0.35	0.19	0.16	0.35	0.19	0.16	0.35	0.19
$q \rightarrow none$	0.16	0.34	0.24	0.16	0.35	0.31	2.57	0.90	1.44
$\alpha \rightarrow none$	0.17	0.35	0.19	0.19	0.35	0.19	0.20	0.35	0.19
$a_z \rightarrow none$	0.46	0.34	0.19	0.45	0.36	0.19	2.77	0.93	1.38
$q \rightarrow \alpha$	0.15	0.34	0.23	0.19	0.36	0.34	2.13	0.71	1.18
$\alpha \rightarrow q$	2.01	0.70	1.03	2.43	1.10	1.29	6.27	2.13	3.22
$q \rightarrow a_z$	0.62	0.47	0.37	2.32	0.76	1.22	2.57	0.90	1.44
$a_z \rightarrow q$	0.90	0.27	0.37	2.66	0.86	1.28	2.76	0.92	1.37
$\alpha \rightarrow a_z$	2.16	0.76	1.02	7.12	2.37	3.55	6.27	2.11	3.15
$a_z \rightarrow \alpha$	0.46	0.35	0.19	0.49	0.36	0.19	0.44	0.36	0.19

Table 6.7 Summary of the fault detection and accommodation results

		Soft additive	Hard additive	Step-type
a	q-NN	1.82	1.16	0.99
	α -NN	2.11	1.28	1.00
	a_z-NN	2.78	1.80	0.80
b	q-NN	0.72	1.62	2.61
	α -NN	0.43	0.72	0.97
	a_z-NN	0.40	0.88	1.38

a: Mean detection time (MT) in seconds
b: Mean fault accommodation errors shown in deg/s, deg and m/s^2 for q-NN, α-NN and a_z -NN respectively.

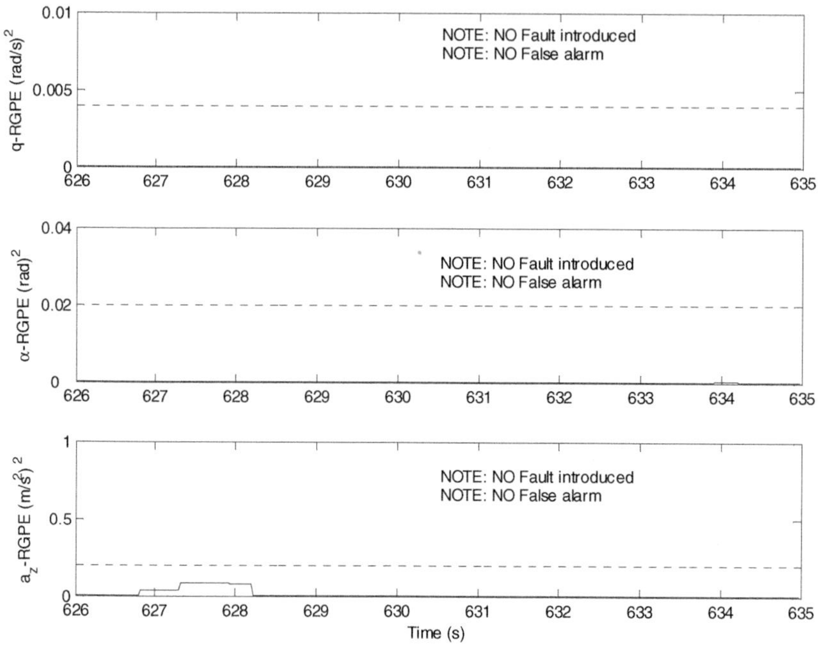

(a) Residuals

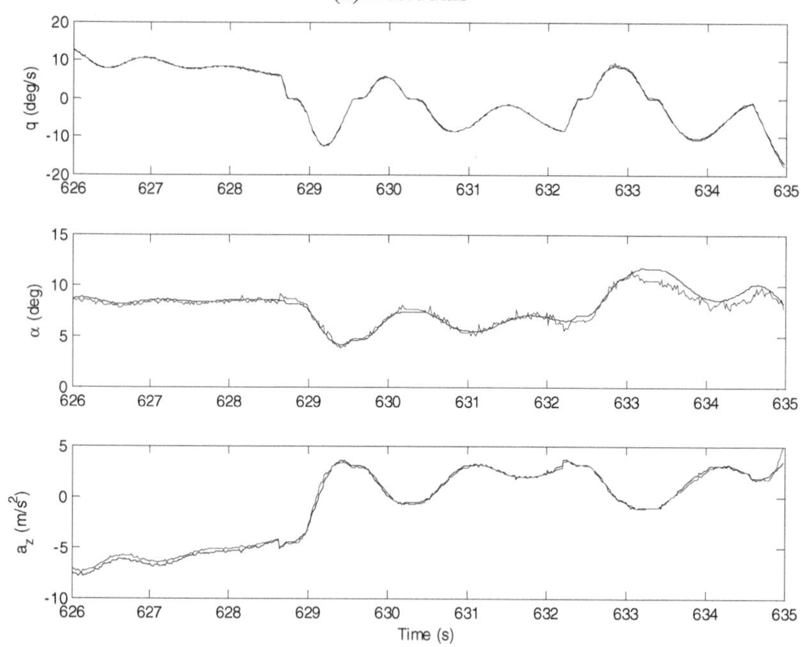

(b) NN estimations

Fig. 6.8 (a) Residual and (b) NN estimations when faults are not present. Note in plot (b): purple (NN), black (ideal).

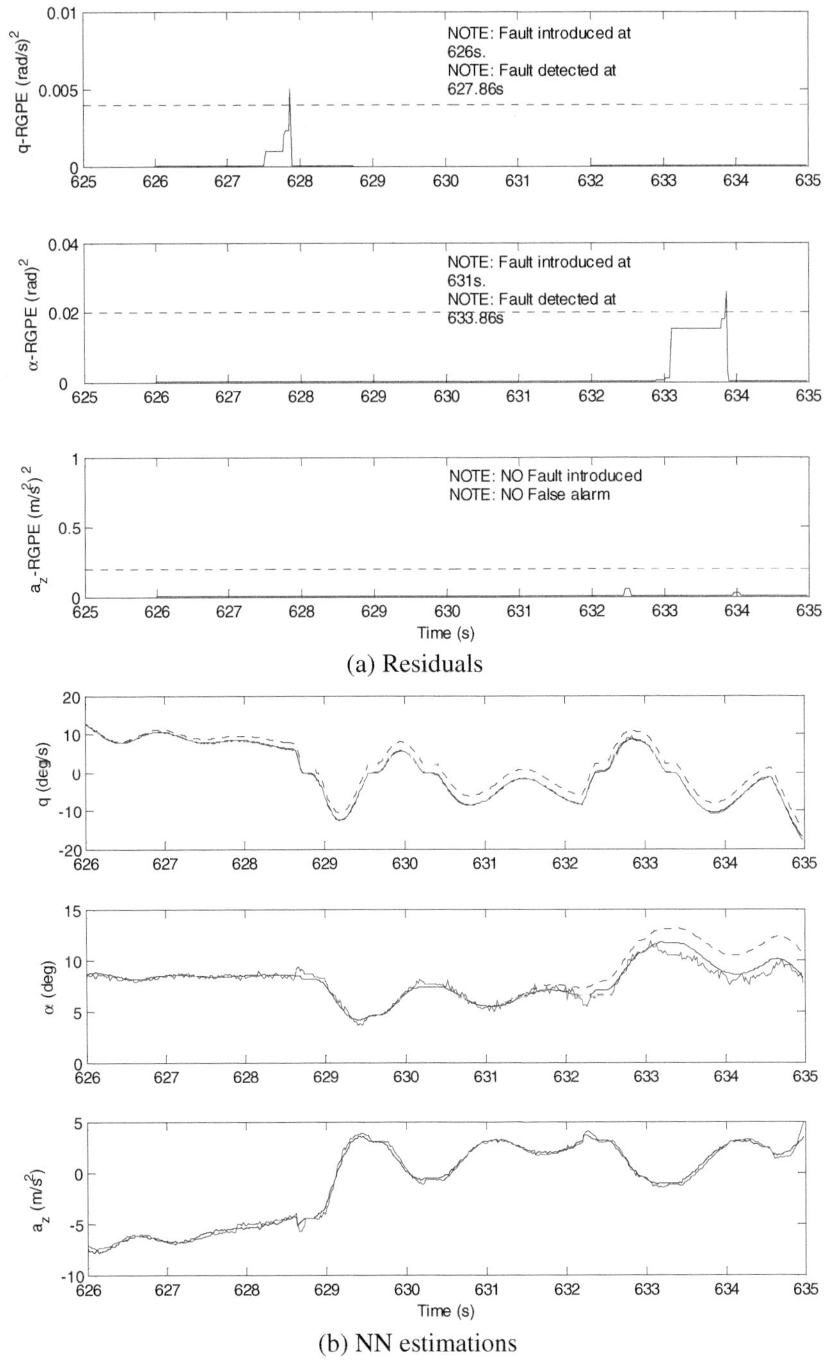

Fig. 6.9 (a) Residual and (b) NN estimations for $\boldsymbol{q} \rightarrow \boldsymbol{\alpha}$ sequence, soft additive fault. Note in plot (b): dotted (faulty), purple (NN), black (ideal).

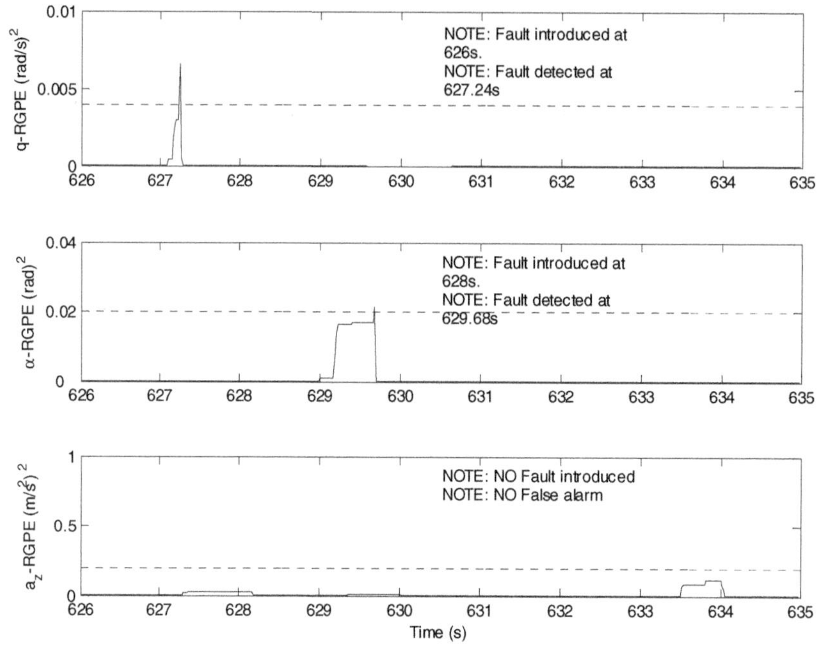

(a) Residuals

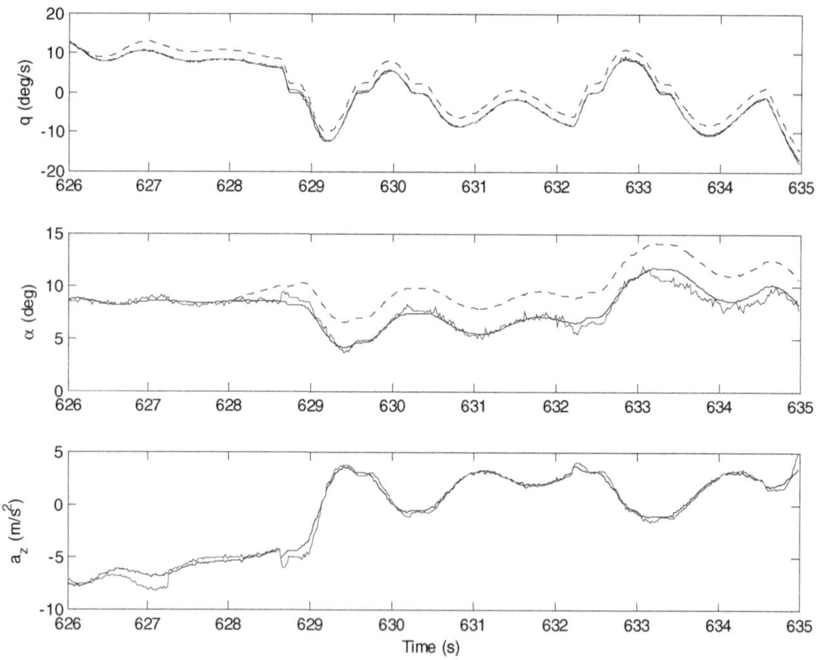

(b) NN estimations

Fig. 6.10 $\boldsymbol{q} \rightarrow \boldsymbol{\alpha}$ hard additive fault. Note in plot (b): dotted (faulty), purple (NN), black (ideal).

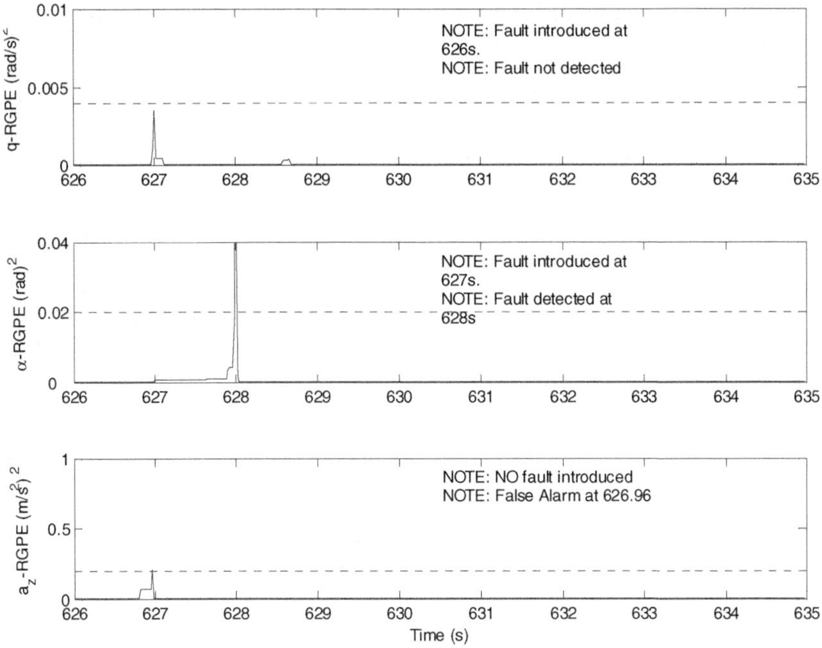

(a) Residuals

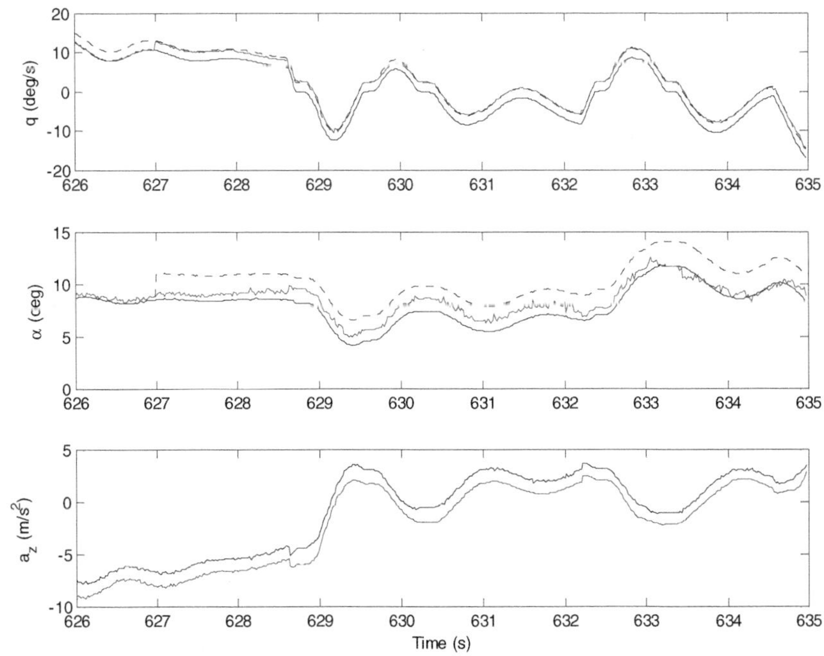

(b) NN estimations

Fig. 6.11 $\boldsymbol{q \rightarrow \alpha}$ step fault. Note in plot (b): dotted (faulty), purple (NN), black (ideal).

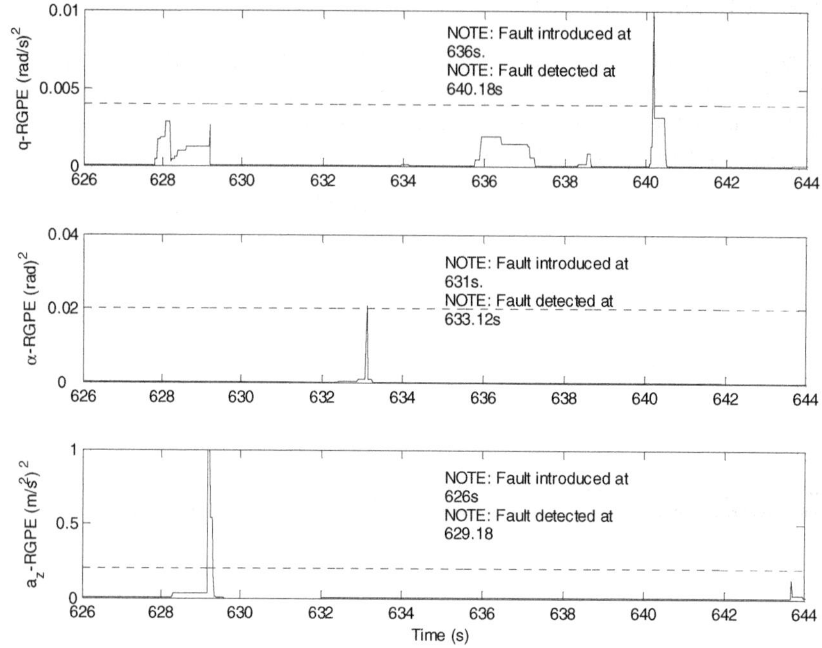

(a) Residuals

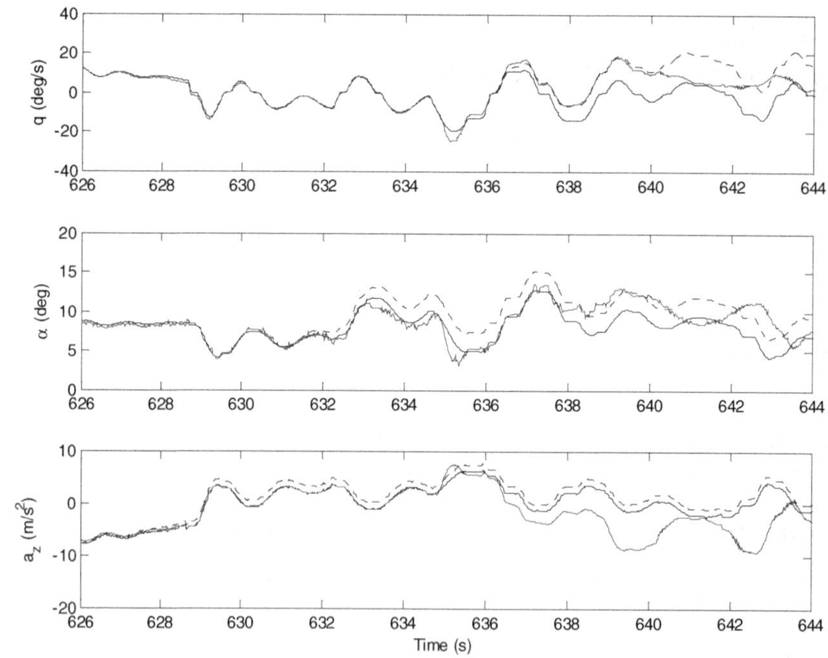

(b) NN estimations

Fig. 6.12 $\boldsymbol{a_z \rightarrow \alpha \rightarrow q}$ soft fault. Note in plot (b): dotted (faulty), purple (NN), black (ideal).

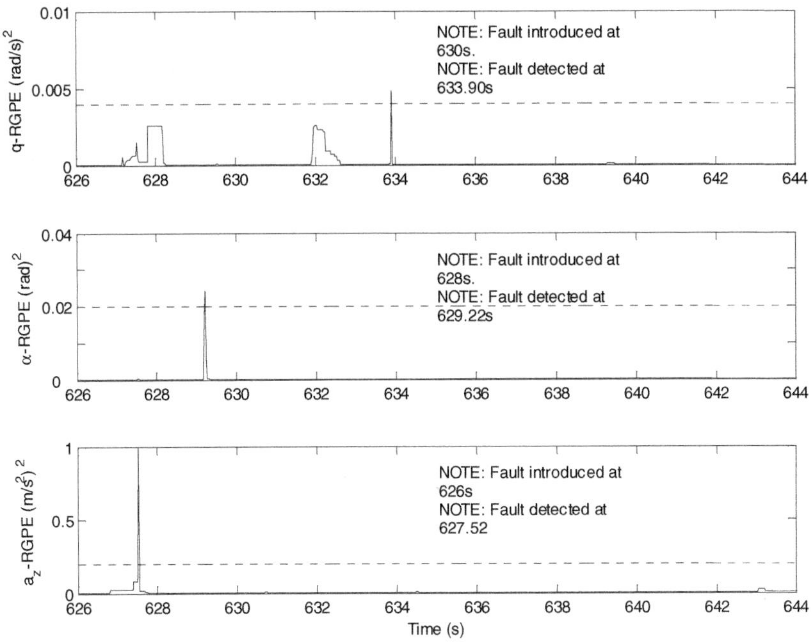

(a) Residuals

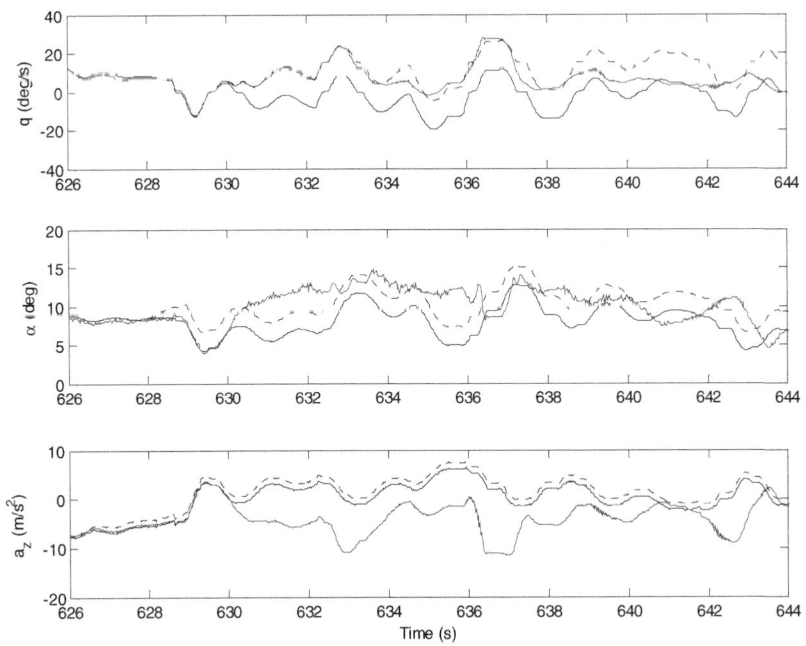

(b) NN estimations

Fig. 6.13 $\boldsymbol{a_z} \rightarrow \boldsymbol{\alpha} \rightarrow \boldsymbol{q}$ hard fault. Note in plot (b): dotted (faulty), purple (NN), black (ideal).

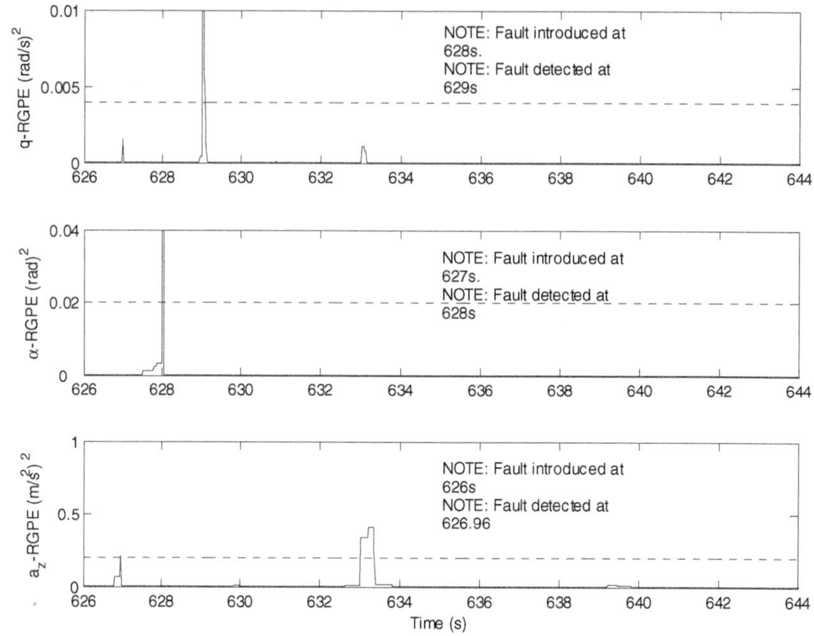

(a) Residuals

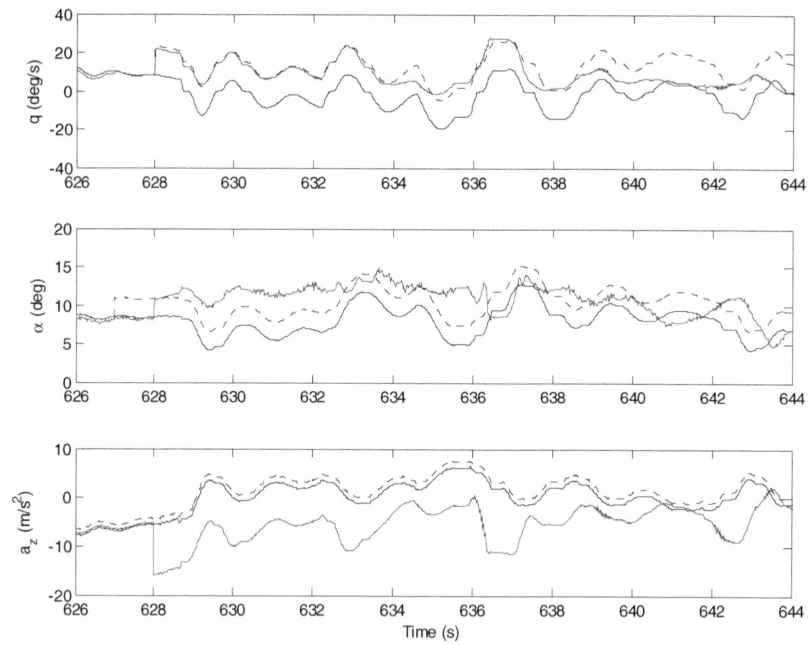

(b) NN estimations

Fig. 6.14 $\boldsymbol{a_z \rightarrow \alpha \rightarrow q}$ step fault. Note in plot (b): dotted (faulty), purple (NN), black (ideal).

Conclusions

In this chapter we design and test a NN-based SFDIA scheme for the detection of multiple sensor faults in a nonlinear UAV model. In this chapter we also extend the single sensor fault scenarios carried out in Chapter 5. The results showed that for 30 separate SFDIA tests, on average; faults were detected in 1.53s, the fault accommodation errors were 1.65 deg/s, 0.71deg, 0.88 m/s^2 for q-NN, α-NN and a_z -NN respectively, 11 false alarms and 2 undetected faults were present overall and the execution time was 0.55ms per data sample.

In general it was concluded that a NN-based SFDIA scheme is highly interconnected especially during the fault accommodation stage, as the output from one NN is used in the input of the other NNs. As a result the performance of the NN-based SFDIA scheme is greatly dependant on the fault detection time. Large fault detection times can result in permanent NN contamination. The RGPE method proposed in Chapter 5 is desirable if we are to reduce the false alarm rates and number of undetected faults. However it also increases the fault detection time which can significantly contaminate interconnected NN models and degrade the overall SFDIA performance. Additionally it was found that whilst step-type faults and constant bias faults can be detected much quicker than hard and soft additive faults, the former can result in permanent NN contamination while the latter generally results in temporary NN contamination. There are several solutions to improve the NN-based SFDIA performance which include tuning the RGPE algorithm (e.g. by reducing the averaging size Ω or number of padding points p_{pad}) so that faults are detected much faster, or the NN can be redesigned to be less sensitive to faults seen in its input set. Examples of the latter include increasing the number of parameters and memory storage used in the NN input set.

Chapter 7
FADS System Applied to a MAV

Introduction

In this chapter a FADS system is designed and wind tunnel tested on a MAV wing (supplied by BlueBear Systems Research, BBSR Ltd). For an introduction to FADS systems the reader is referred to Chapter 3. The aim in this chapter is to design a FADS system which can convert aircraft surface pressure to 3 air data states; $P_\infty, V_\infty, \alpha$. The ideal location of the pressure ports (orifices) on the MAV wing must be first investigated via CFD simulation to avoid placing the ports at areas where surface pressure is insensitive to the desired air data. In our case both 2D and 3D CFD simulations are implemented, wind tunnel tests are carried out at Leicester University and the flight data is processed (offline) using an EMRAN RBF NN (Chapter 4).

This chapter is organised as follows. The MAV is presented in section 7.1. Section 7.2 discusses the 2D CFD simulations applied to a section of the MAV wing (i.e. an aerofoil) and the results are used to conclude a suitable location of the pressure ports. The final pressure port locations are outlined in section 7.3. In section 7.4 a 3D CFD simulation of the wing is also carried out to explain how the sideslip (β) cannot be accurately estimated based on the pressure port locations chosen. Sections 7.5-7.7 discuss the wind tunnel tests and section 7.8 is the results section where the FADS system is analysed in terms of; estimation accuracies for wind tunnel static and dynamic tests, fault tolerance performance and performance in comparison to a standard lookup table (LUT) approach.

7.1 The Mini Air Vehicle (MAV)

The MAV used to test the FADS system is shown in Fig 7.1. The MAV uses a MH64 wing section [144] and flies at a maximum speed of 20m/s. From Fig 7.1 we can see that the MAV is driven by a nose propeller and therefore mounting a FADS system at the wing leading edge is more suitable than the nose. Furthermore the wing leading cannot be approximated as a sphere and therefore a NN modelling approach in the FADS system, is more appropriate than the aerodynamic model (which is derived based on spherical shape assumptions) defined in Chapter 3, section 3.3. Some of the MAV properties are shown in Table 7.1. However note that only the wing section (Fig 7.2) is used in the wind tunnel and CFD tests.

I. Samy and D.-W. Gu: Fault Detection and Flight Data Measurement, LNCIS 419, pp. 109–158.
springerlink.com

Fig. 7.1 MAV (courtesy of BBSR Ltd.)

Table 7.1 Some of the MAV characteristics

Characteristic	Value
Speed range	8-20 m/s
Mass	450 g
Wing span	488 mm
Wing root chord	250 mm
Wing tip chord	200 mm
Wing thickness (t/c)	8.61%

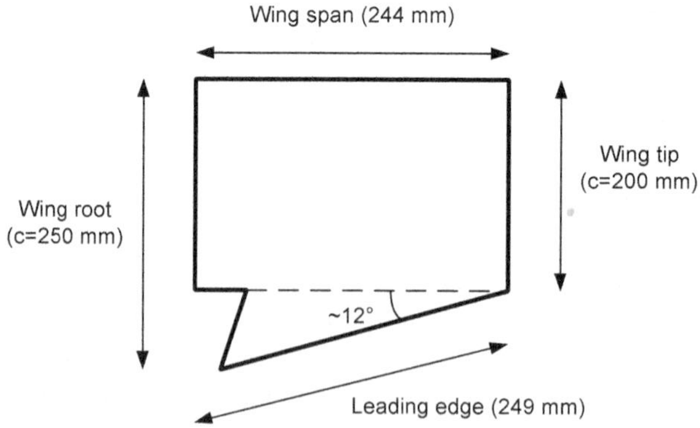

Fig. 7.2 Top view of the wing section (c is the wing chord)

7.2 CFD Simulations (2D)

CFD simulations were performed on the MAV wing using the MH64-wing coordinates defined in [144]. In general it is useful to perform CFD simulations prior to the actual tests as they allow us to tune and test our designs with minimal incurred costs.

A CFD analysis of the pressure distribution over a wing can help us identify which parts of the wing would be most suitable for mounting the pressure ports in the FADS system. The purpose of a FADS system is to convert surface pressure to air data states, and it is therefore important that the pressure ports are mounted in areas which are highly sensitive to the air data. In our case the air data only includes; $P_\infty, V_\infty, \alpha$.

7.2.1 *Background and Terminologies*

An aerofoil is any section of the wing cut by a plane perpendicular to the wing (Fig 7.3). It is 2D and can be normalised (i.e. chord length is set to 0-1) to investigate the general aerodynamic properties of the wing (i.e. lift, drag, pressure distribution etc.). 2D CFD simulations of an aerofoil can have some limitations. For example we are unable to investigate the aerodynamic properties vs. sideslip. In more advanced applications, a 2D aerofoil would be unsuitable if the wing tip vortices are of interest to the engineers. However in our case we are only interested in *approximating* the pressure distributions in order to locate the steep pressure gradients at the wing leading edge. Therefore as we are not interested in simulating the exact air flow over the wing and calculating the exact pressure magnitudes, a 2D CFD simulation is suitable and more importantly strict convergence criteria are not necessary.

The air flow simulated over an aerofoil and indeed the whole aircraft can be categorised according to; its *viscosity* and *compressibility* [124]. Air flowing past an aerofoil is essentially *viscous*. Viscosity effects are caused by the friction between the air and the aerofoil, the thermal conduction between areas of high and low temperature and the mass diffusion when fluid concentration gradients are present [124]. Air is also *compressible* which implies that it would consistently change its density (ρ) as it expands and contracts throughout flow. However for most low speed flights (speeds less than $\approx$ 100 m/s), it can be assumed that the air flow is inviscid and incompressible [145]. This is mathematically convenient and can simplify the CFD simulation. In theory however, such a flow does not exist. For example the Reynolds number (Re) is a dimensionless number which is essentially the ratio of inertial forces to viscous forces. Therefore an inviscid flow requires an infinite Re which cannot exist in reality. Furthermore the frictional forces of the viscous air flow are a major contribution to the aerodynamic drag and therefore assuming inviscid flow cannot on its own predict the total drag.

Assuming inviscid, incompressible flow, a well-known equation can be used to relate the pressure (p) and airspeed (V) anywhere in the flow field [124]:

$$p + \frac{1}{2}\rho V^2 = \text{constant} \tag{7.1}$$

Equation 7.1 is the well-known Bernoulli equation were p is the local pressure, ρ is the (constant) air density and V is the airspeed. The Bernoulli equation states that at any point in the air flow, if the velocity increases then the local pressure at that point decreases and vice versa. It is important to note that (7.1) would only be applicable in inviscid, incompressible flows where there is no energy dissipation due to viscous effects (e.g. friction) and the air density is constant.

There are several useful parameters and relationships that can be derived from (7.1). These include the pressure coefficient (Cp) and the relationship between total pressure (P_0), freestream static pressure (P_∞) and dynamic pressure (q_∞). Using (7.1) we can define the relationship between any two points (1 and 2) in the air flow as:

$$p_1 + \frac{1}{2}\rho V_1^2 = p_2 + \frac{1}{2}\rho V_2^2 \tag{7.2}$$

As defined in Chapter 3 (section 3.1), P_0 is the pressure measured when $V \approx 0$. Therefore if point 1 in (7.2) is taken to be at P_0 then:

$$P_0 + \frac{1}{2}\rho (0)^2 = p_2 + \frac{1}{2}\rho V_2^2 \tag{7.3}$$

$$P_0 = p_2 + \frac{1}{2}\rho V_2^2 \tag{7.4}$$

If point 2 in (7.4) is measured far ahead from any flow disruptions (due to the aerofoil) then any aerodynamic property measured at that point is referred to as *freestream*. For example the airspeed measured at the surface of an aerofoil would not be freestream as the air would be either accelerated or decelerated as it collides with the surface of the aerofoil. Therefore referring back to (7.4), if point 2 is measured far ahead of the aerofoil then:

$$P_0 = P_\infty + \frac{1}{2}\rho_\infty V_\infty^2 \tag{7.5}$$

where subscript ∞ indicates freestream and:

$$q_\infty = \frac{1}{2}\rho_\infty V_\infty^2 \tag{7.6}$$

We can also define the pressure coefficient (Cp) as [124]:

$$C_p = \frac{p - p_\infty}{q_\infty} \tag{7.7}$$

where p is the local pressure. Equations (7.5)-(7.7) are useful when analysing the air flow in CFD simulations as well as real applications. For example a Pitot-static tube measures P_0 and P_∞, and therefore using (7.5), V_∞ can be calculated. However note that the airspeed calculated using (7.5) would be the indicated airspeed, i.e. we are assuming $\rho = \rho_\infty$ everywhere (incompressible flow). Cp is also an important parameter as most pressure distribution curves found in the aerodynamic literature are defined in terms of this dimensionless number instead of the pressure magnitude.

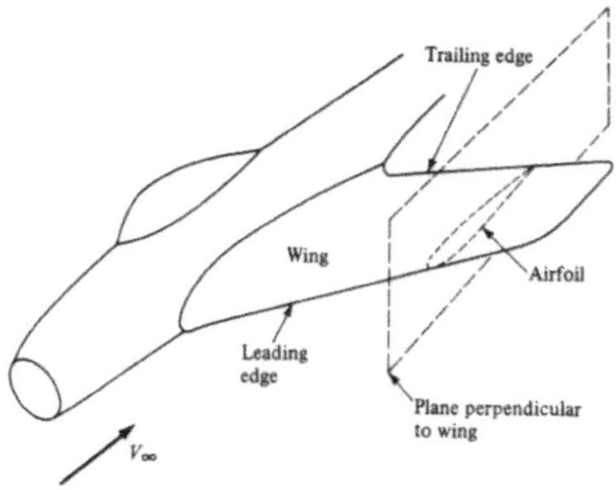

Fig. 7.3 Aircraft terminologies [121].

7.2.2 Results

The software package Gambit is used to build the MH64 aerofoil which is later analysed in the CFD software, Fluent. Fig 7.4 shows the aerofoil build in Gambit. The quadrilateral 'cells' are generated during the meshing stage. The aerodynamic properties anywhere in a cell are constant. Meshing is the process where a large domain is subdivided into small sections and each section is then analysed separately. This reduces the modelling complexities. Increasing the number of cells can increase the accuracy of the CFD simulations but it can also increase the processing time. In our case, the domain around the aerofoil was divided into 12240 cells (Fig 7.4).

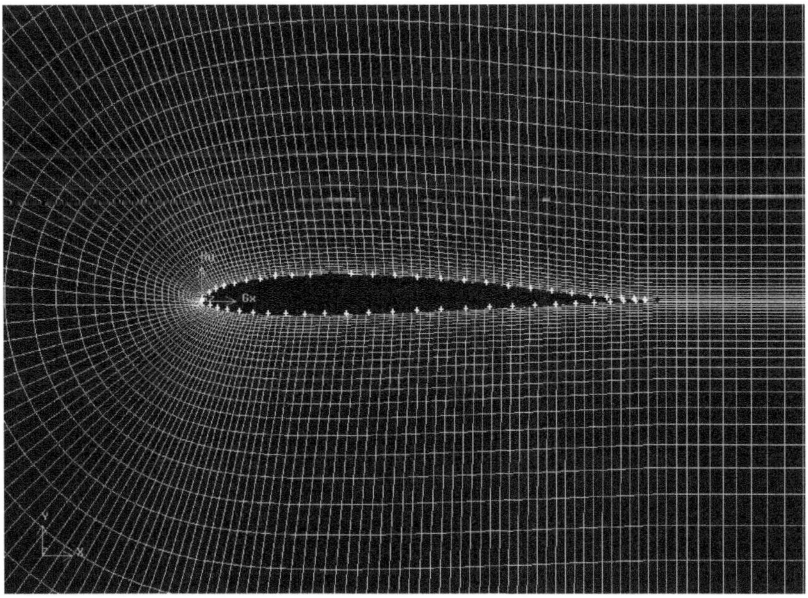

Fig. 7.4 Gambit 2D, 12240 quadrilateral cells

The aerofoil designed in Gambit can then be uploaded into Fluent. In our case the air flow was assumed to be inviscid and incompressible. Standard sea-level atmospheric conditions were also considered and pressures were referenced to the atmospheric pressure (101.325 kPa). The convergence criterion was set to 1e-5 for all CFD simulations, and tests were carried out at V_∞ =12, 15, 18, 20 m/s and α= 0, 2, 4, 6, 8 deg.

Figs 7.5 and 7.6 show an example of the velocity and pressure distributions respectively, for V_∞=20m/s and α=0deg. From Fig 7.5 we can see that the airspeed far away from the aerofoil is almost 20 m/s. This is expected, as the freestream airspeed is in fact, V_∞ = 20m/s. As the air approaches the aerofoil, the airspeed changes due to the collision of the air with the aerofoil surface. So for example we can see from Fig 7.5 that the flow is accelerated over the upper surface of the aerofoil (as it is pushed along the surface). On the other hand at the leading edge we will find that the airspeed is at its minimum. This is in fact the stagnation area of the aerofoil where the air is decelerated to approximately zero airspeed and is also where the total pressure P_0 can be measured.

Fig 7.6 can be related to Fig 7.5 via the Bernoulli equation (7.1). For example we will notice that at areas where the airspeed is accelerated ($V > V_\infty$ in Fig 7.5), the corresponding pressure decreases ($P < P_\infty$ in Fig 7.6). If we carefully observe Fig 7.6 we will also notice that despite both being negative, the pressure at the upper surface of the aerofoil is lower than the pressure on the lower surface. This net pressure imbalance is in fact what gives the lifting property of aircraft wings.

In our investigation we are interested in locating the pressure gradients on the aerofoil surface. For this purpose we require an x-y plot of aerofoil surface pressure vs. chord. As discussed in section 7.2.1, it is more convenient to plot Cp (7.7) instead of the pressure magnitude. The Cp plots are displayed in Figs 7.7-7.10. To better understand the Cp plots let us re-define Cp in (7.7) in terms of the ratio of local airspeed (V) to freestream airspeed (V_∞):

$$C_p = 1 - \left(\frac{V}{V_\infty}\right)^2 \tag{7.8}$$

Note that (7.8) is only applicable in inviscid, incompressible flows [124]. From (7.8) we can arrive at the following:

- Cp~1 at the stagnation point (i.e. where V~0).
- Cp<1 if the flow is accelerated (i.e. $V > V_\infty$).
- Cp>1 if the flow is decelerated (i.e. $V < V_\infty$).

The pressure ports of the FADS system must be mounted on the wing in areas where:

1. There are steep pressure gradients
2. There are strong variations between the upper and lower surface pressures.

This offers a wealth of non-redundant information from which the air data values may be estimated [38]. From Fig 7.7 we can see that criteria 1 and 2 are satisfied for x/c < 0.1. If we zoom into Fig 7.7 (Fig 7.8) we can conclude that most of the 'action' occurs at x/c < 0.01 and Cp is almost constant for x/c > 0.01. Therefore initially we can conclude that the pressure ports must be located at x/c < 0.01.

We must next consider the variation of pressure with V_∞. However in this case Cp would not change for a change in V_∞. This is because the ratio V/V_∞ would remain the same and therefore Cp in (7.8) would be constant. Instead we consider the static pressure on the surface of the aerofoil (Fig 7.9-7.10). Once again from Fig 7.10 we find that the steep pressure gradients occur at x/c < 0.01.

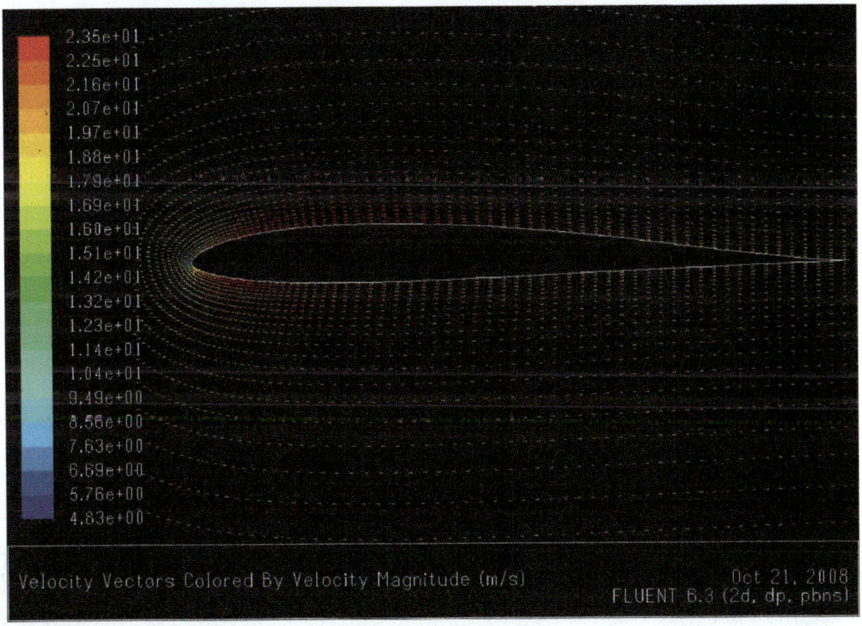

Fig. 7.5 Velocity distribution for $\boldsymbol{V}_\infty$ = 20m/s, $\boldsymbol{\alpha}$ = 0deg

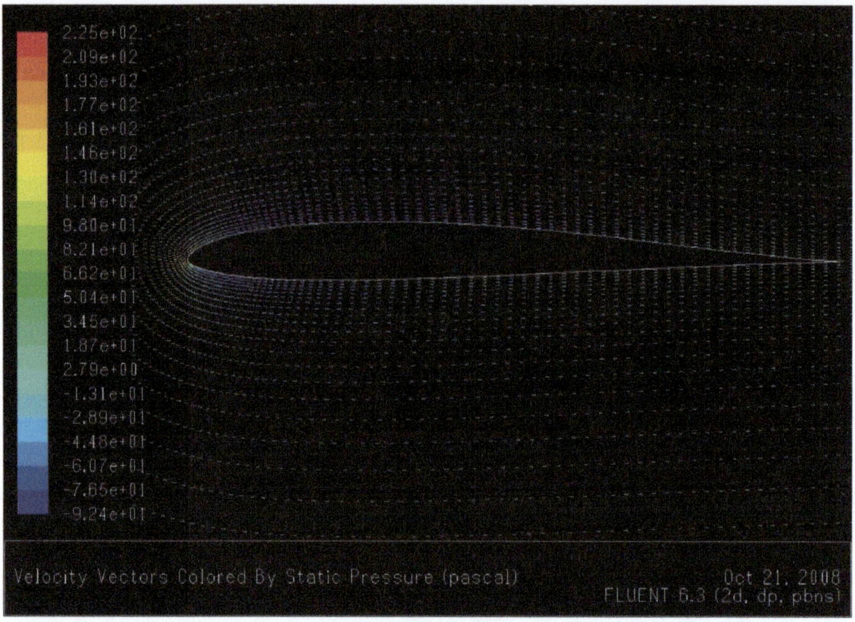

Fig. 7.6 Pressure distribution for $\boldsymbol{V}_{\infty}$ = 20m/s, $\boldsymbol{\alpha}$ = 0deg

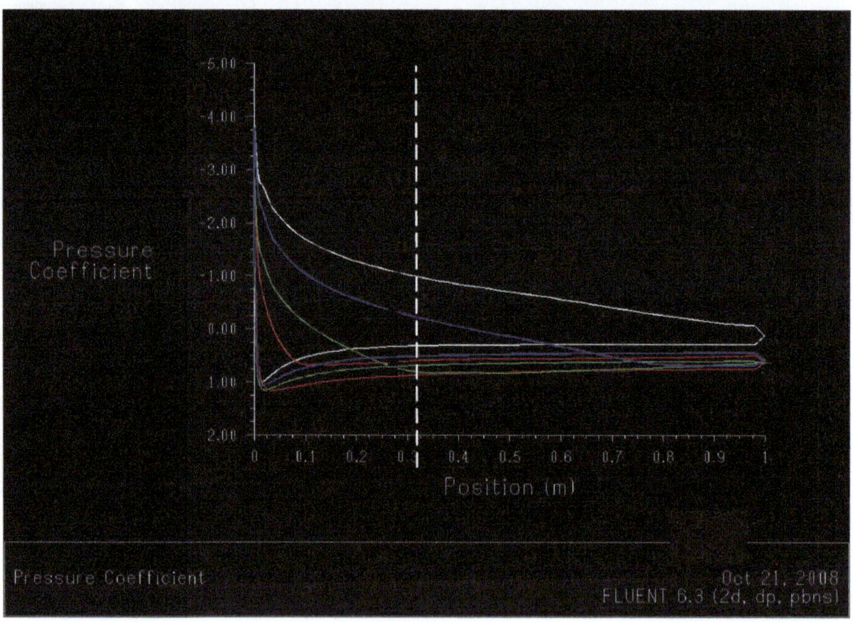

Fig. 7.7 Cp plot for $\boldsymbol{V}_{\infty}$ = 20m/s, $\boldsymbol{\alpha}$ = 2, 4, 6, 8 degs

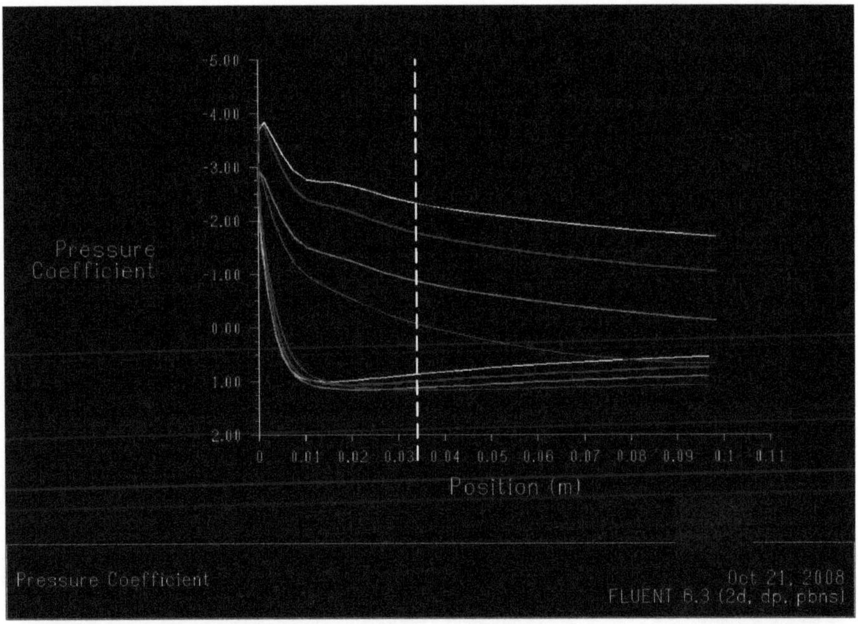

Fig. 7.8 Zooming into Fig 7.7

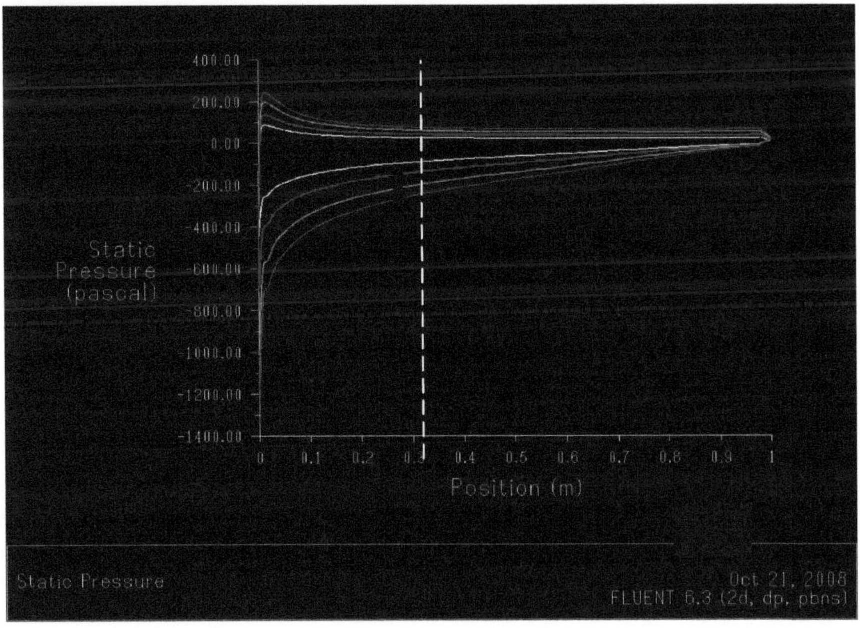

Fig. 7.9 Cp plot for $\boldsymbol{V}_{\infty}$ = 12, 15, 18, 20m/s and $\boldsymbol{\alpha}$ = 2 deg

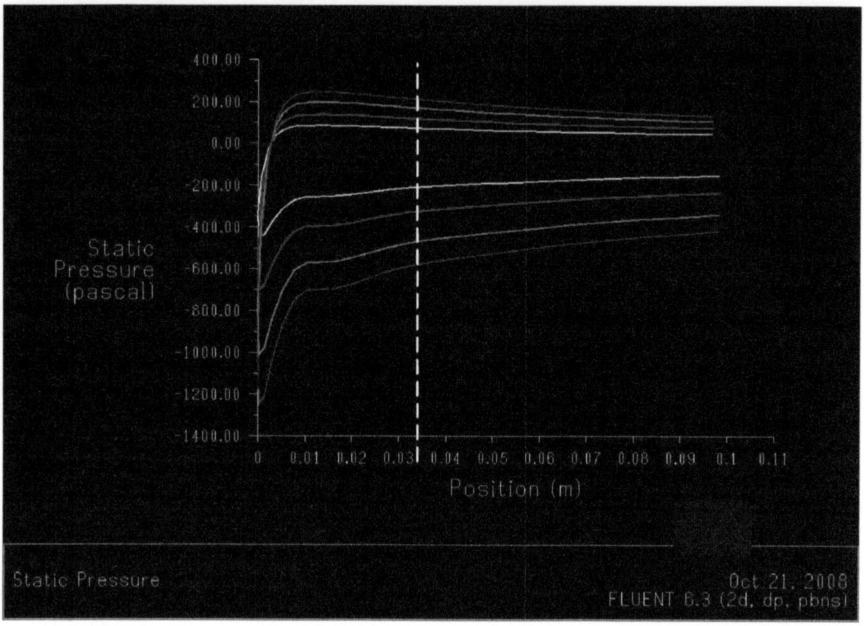

Fig. 7.10 Zooming into Fig 7.9

7.3 Location of the Matrix of Pressure Orifices (MPO)

The number of pressure ports (orifices) must be chosen as a compromise between the need to accurately estimate the air data states and the need to reduce instrumentation costs. Choosing a large number of pressure ports can improve the fault tolerance properties of the FADS system and its robustness to noise caused by e.g. air turbulence or instrumentation noise. However not only do a large number of pressure ports increase the instrumentation cost, but in MAVs we must take into account the space and weight limitations. In our case there are 3 air data states to be estimated ($P_\infty, V_\infty, \alpha$) and therefore a minimum of three pressure ports are required. Two extra ports are added in order to improve the redundancy options and the noise sensitivity of the FADS system. We will from now on refer to these five pressure ports collectively as the matrix of pressure orifices (MPO) which include P1, P2, P3, P4, and P5 (Fig 7.11).

It is important that the MPO are distributed in such a way that 1) ports close to each other give different pressure measurements (this avoids having redundant information) and 2) the pressures measured are sensitive to variations in the air data states $P_\infty, V_\infty, \alpha$. In section 7.2.2 we concluded that the latter criterion is satisfied at x/c <0.01. Taking this into account, two pressure ports were placed on the upper surface of the wing, two on the lower surface and one close to the tip of the wing leading edge (see 'Side-View' of Fig 7.11). The latter port is included as a stagnation pressure source (i.e. for approximate measurements of P_0).

As far as the author is aware, there is no universally accepted standard for the minimum hole spacing (i.e. the distance between adjacent pressure ports) and indeed the size of the pressure ports (i.e. the diameter of the pressure orifices). For example [146], [32], [147], [43] use a hole diameter of 0.76, 0.79, 6.40, 5.00 mm and a minimum hole spacing of 1.08, 15.25, 61, 406 mm respectively. In general it was found from the literature that a minimum hole spacing of 1mm and a hole diameter of 0.5mm were suitable.

Distributing the pressure ports along the wing span (see 'Front-View' of Fig 7.11) can help increase the hole spacing. This is important, for two reasons. Firstly this avoids having redundant information. Secondly it can avoid any flow interference between the adjacent ports. Pressure orifices drilled into the surface of the wing will result in uneven edges around the port and so the air flow over the orifices will not be smooth. This disruption of the air flow can consequently affect the pressure measurements of adjacent ports. This flow interference effect between adjacent ports is one of the main reasons why pressure ports in FADS systems must be suitably spaced.

Consider the 'Side-View' of Fig 7.11. To increase the hole spacing we could of course spread the ports over the whole aerofoil (i.e. shifting the pressure ports towards the trailing edge). However, by doing so we would no longer fulfil the criterion that ports must be placed at $x/c<0.01$. Therefore instead we distribute them along the wing span (see 'Front-View' of Fig 7.11).

In conclusion there are three important criteria when choosing the locations of the 5 pressure ports:

1. They must all be located at $x/c < 0.01$.
2. A minimum hole spacing of 1mm and hole diameter of 0.5 mm must be used.
3. The MPO must be placed far enough from the wing root and tip.

The latter criterion is important if we are to avoid the effects of flow separation at the wing tip and the turbulent air behind the nose propeller, i.e. the nose propeller wash. For this reason the middle port P3 (Fig 7.11) was placed at a span of 200mm from the wing root (i.e. 44mm from the wing tip). Table 7.2 shows the locations of the 5 pressure ports in terms of x/c and Fig 7.11 shows the MPO mounted on the MAV wing where 0.50 mm hole diameter and 5.00 mm hole spacing were used.

Table 7.2 Pressure port locations (also refer to Fig 7.11)

Port	x/c position
P1	0.009
P2	0.003
P3	0.001
P4	0.003
P5	0.009

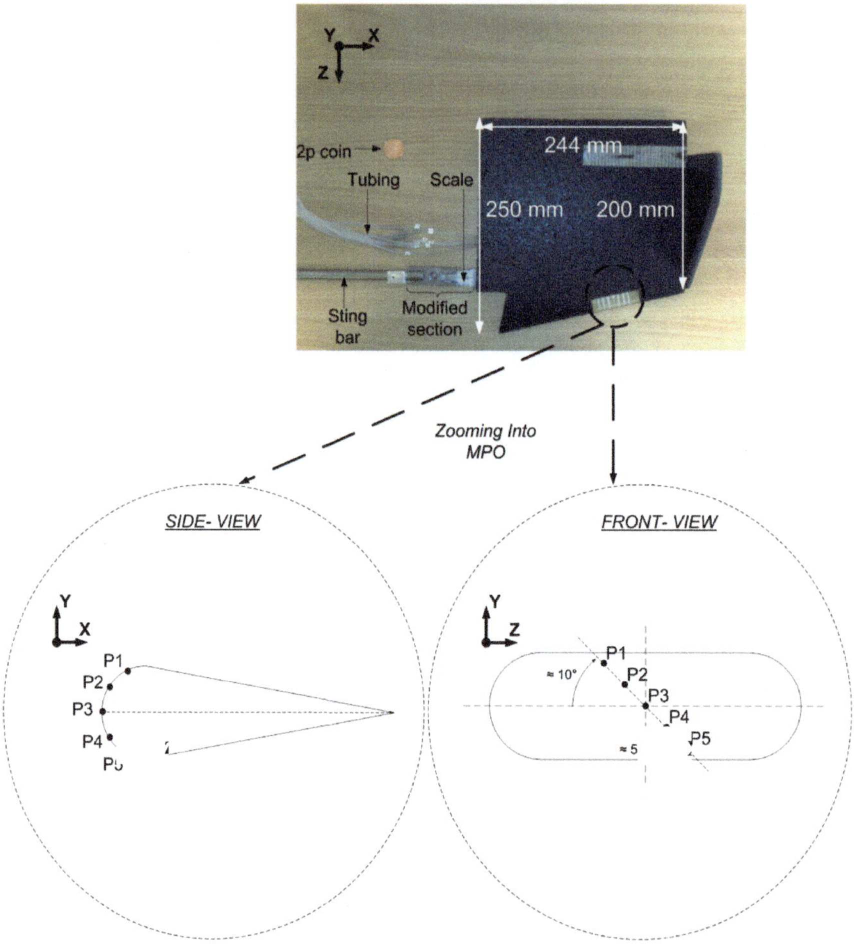

Fig. 7.11 Top view of wing and MPO (with pressure ports P1, P2, P3, P4, P5), MPO not shown to scale

7.4 CFD Simulations (3D)

The locations of the five pressure ports (P1, P2, P3, P4, P5) have been chosen so that 3 air data states $(P_\infty, V_\infty, \alpha)$ can be estimated from the FADS system. Therefore a 2D CFD simulation was sufficient to conclude the areas of steep pressure gradients at the wing leading edge (section 7.2-7.3). However we have not yet considered the possibility of estimating the sideslip, β. To investigate this further, a 3D CFD simulation of the whole wing must be implemented, with constant V_∞ and α and changing β.

3D CFD simulations of a body can be accurately implemented only if the exact body coordinates (dimensions) are known. Fortunately, the required simulation accuracy is highly dependent on the application and in our case, as we are mainly interested in locating the steep pressure gradients and not estimating the exact pressure magnitudes, the geometrical model built in Gambit can be greatly simplified.

Three assumptions were considered when building the wing in Gambit (compare the Gambit model in Fig 7.12 with the real wing in Fig 7.11):

1. The winglet at the wing tip is not included
2. The indent at the leading edge close to the wing root (Fig 7.11) is not included
3. The wing thickness was assumed constant across the wing span

As designed in section 7.3, the MPO of the FADS system is located far away from the wing root and wing tip. Therefore assumptions 1 and 2 above should not significantly affect the CFD results. The final assumption is quite extreme and can result in inaccurate CFD simulations. However we should still be able to roughly identify the pressure gradients vs. sideslip. Furthermore the MPO (Fig 7.11) is confined to a small section on the wing leading edge and therefore the change in thickness, and indeed the chord length, within this small section is not large. The wing built in Gambit and the meshed grid is shown in Fig 7.12 and Figs 7.13-7.14 respectively. In comparison to the 2D aerofoil (Fig 7.4, section 7.2.2), 1042950 hexahedral cells were used. As a result the 3D CFD simulations were much slower with one test taking over six hours to complete.

Fig. 7.12 3D-wing section built in Gambit

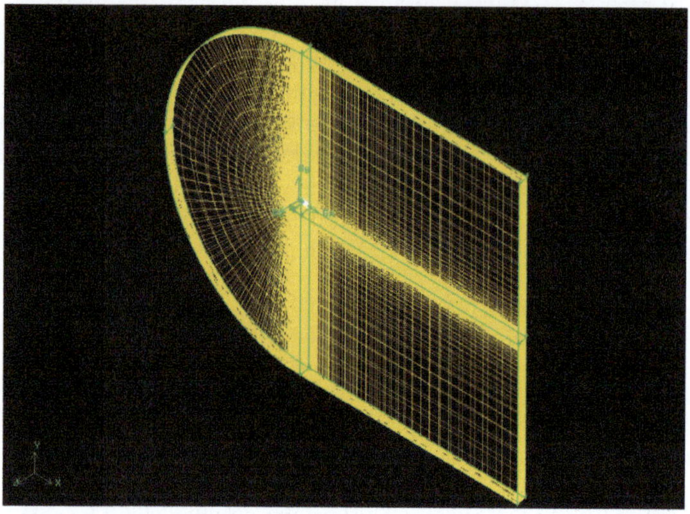

Fig. 7.13 3D-wing and meshed grid built in Gambit (meshed with 1042950 hexahedral cells)

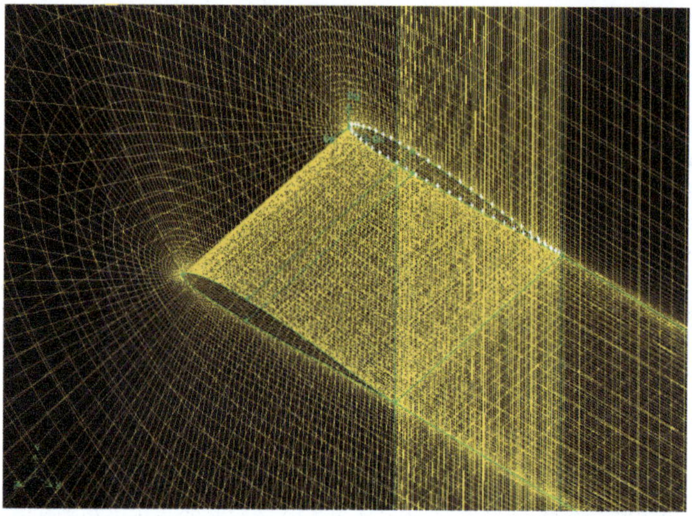

Fig. 7.14 Zooming into Fig 7.13

As in the 2D CFD simulations, the air flow was assumed inviscid and incompressible. The tests were carried out in Fluent at a fixed V_{∞}=15m/s and α=0deg, and only the sideslip was varied between β=0-10deg. The 3D CFD simulation results are shown in Figs 7.15-7.17.

The velocity profile of the wing is shown in Fig 7.15. As we earlier observed in Fig 7.5, we notice that the airflow is decelerated as it collides with the wing leading edge ($V < V_\infty$) and then accelerated ($V > V_\infty$) over the upper surface of the wing. Figs 7.16-7.17 are the Cp plots vs. β and can be used to identify which parts of the wing are sensitive to changes in sideslip.

Let us first consider Fig 7.16. There are 84 separate lines representing different parts of the *wing leading edge*. Note that the wing leading edge is assumed to be a line at the front end of the wing (see Fig 7.2). This line is divided into 84 sections and the pressure is measured separately at each section. The Cp plot for each section is shown in Fig 7.16. So for example in Fig 7.16, the line at the top of the plot represents the section at the wing root while the bottom line represents the section at the wing tip. The length of the leading edge is 249 mm (see Fig 7.2) and therefore each section is 2.96mm wide (249mm/84). From Fig 7.16 we notice that Cp varies almost linearly with β for areas close to the wing root and tip. On the other hand in areas far away from the wing root and wing tip, Cp is almost constant for $\beta > 3$deg. Therefore ideally we would want to place the pressure ports closer to the wing root and wing tip due to the almost linear relationship of Cp and β. From Fig 7.16 we can calculate that for parts of the wing leading edge which are between 0-30mm and 225-249mm (represented by a tick) there are almost linear variations of the pressure with the sideslip. Therefore ideally we would want to place the pressure ports in the areas where the tick is marked in Fig 7.16.

Again, Fig 7.17 investigates the variation of Cp with β but this time we move backwards from the wing leading edge i.e. moving towards the trailing edge. In Fig 7.17 we have shifted backwards from the wing leading edge (x/c= 0.006) and again as in Fig 7.16, plotted Cp vs. β for 84 different sections of the wing span. As we can see the changes in Cp vs β are very small. So for example, at the wing tip (top line), Cp only changes by approximately 0.025 between β=0deg and β=10deg. Therefore we can conclude that the further back we go from the wing leading edge the less sensitive the pressure ports would be with sideslip.

In conclusion to estimate the sideslip, the ideal location of the pressure ports is:

1. Close to the wing leading edge
2. In areas marked by a tick in Fig 7.16, i.e. close to the wing root and wing tip.

The first criterion is expected as we have already seen in the 2D CFD simulations (section 7.2) that most of the steep pressure gradients occur at the wing leading edge. The problem with placing the pressure ports close to the wing root and wing tip is that there can be significant disruptions to the air flow caused by the nose-propeller wash and the wing tip vortices respectively. It is for this reason that the MPO was located far away (P5 was 200mm from the wing root, Fig 7.11) from the wing root and wing tip at the cost of not being able to accurately estimate the sideslip, β. Further improvements to the design proposed here will be discussed in the Future Work chapter (Chapter 8).

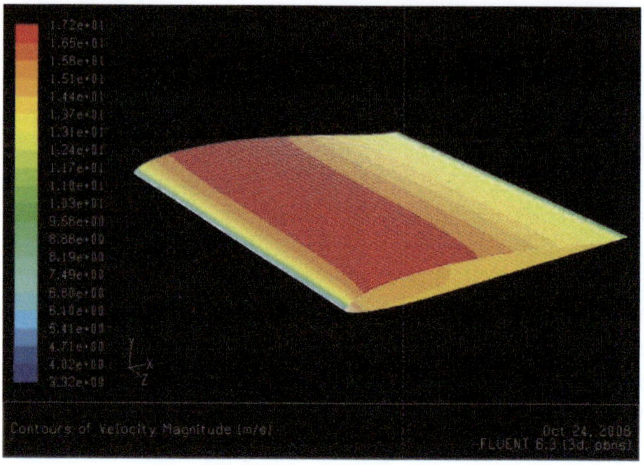

Fig. 7.15 Velocity profile of 3D-wing. $\boldsymbol{V}_{\infty}$ = 15m/s, $\boldsymbol{\alpha}$=0deg and $\boldsymbol{\beta}$=0deg

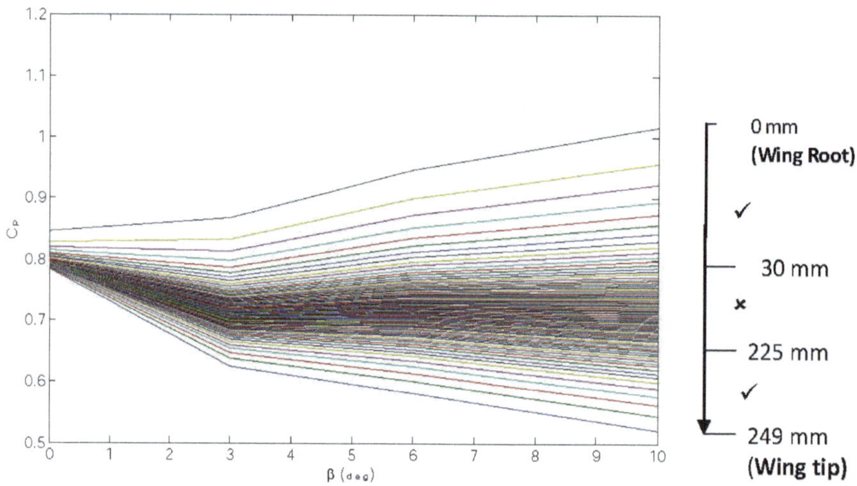

Fig. 7.16 Change in pressure coefficient vs sideslip for each part of the wing leading edge. Increments are in approx 3mm starting from wing root to wing tip. $\boldsymbol{V}_{\infty}$ = 15m/s, $\boldsymbol{\alpha}$=0deg.

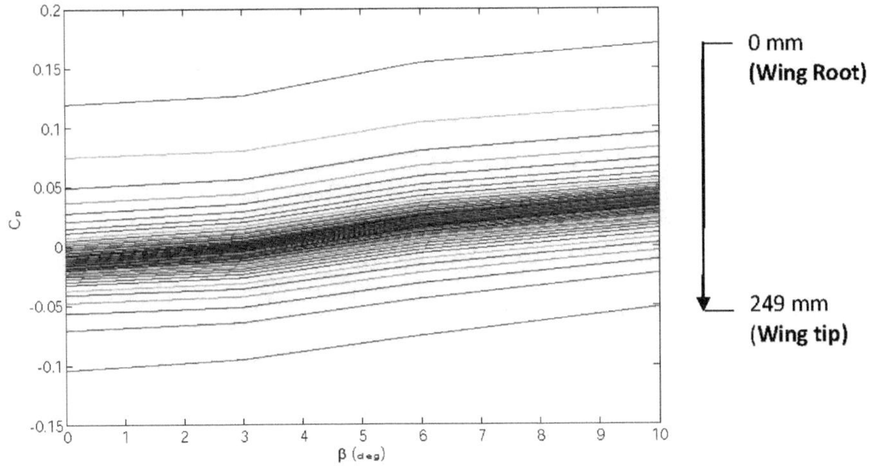

Fig. 7.17 As in Fig 7.16 but moving backwards from the wing leading edge at x/c=0.006.

7.5 Wind Tunnel and Instrumentation

The MAV wing mounted with the FADS system was wind tunnel tested at Leicester University. The wind tunnel is an open-ended subsonic wind tunnel capable of reaching speeds of 40 m/s and has a working section of 0.46m x 0.46m (see Fig 7.18 and Fig 7.19). The MAV wing is sting mounted in the wind tunnel and an external *balance* is used to support the wing via the sting bar. The sting bar is connected close to the wing centre of gravity so that for zero angle of attack settings (set from the balance), the wing would also be at zero angle of attack. The balance supporting the sting bar included a scaled-turning knob (calibrated in deg) which allowed variations in wing α (Fig 7.19). To allow for sideslip variations the standard sting bar was slightly modified (see 'Modified section' in Fig 7.11). The part allowing the sideslip comprises two rectangular pieces of metal with one end rounded to a semicircle. The straight edge of one piece is connected to the sting bar and the straight edge of the other is connected to a bar fixed to the wing. The two plates can rotate about a hole through which passes a bolt to lock them together at the chosen angle. The rounded ends carry a scale to indicate the wing β. Wind tunnel instrumentation includes:

1. Pitot-static tube
2. 6 pressure transducers
3. Potentiometer
4. Data acquisition (DAQ) card
5. PC with LabView installed

A Pitot-static tube was mounted in the wind tunnel ahead of the wing and connected to a differential pressure transducer to measure the difference between P_0 and P_∞. Using this pressure difference, we can calculate V_∞ via (7.5) (assuming inviscid, incompressible flow). The remaining 5 pressure transducers are connected via pressure tubing to the 5 pressure orifices (0.50mm diameter each) where the pressure measured is differential (i.e. with respect to room pressure). It is important to keep the tubing distance to a minimum to avoid any time lags. The pressure transducer board was therefore placed as close as possible to the wind tunnel giving an approximate tubing length of 0.38 m. The pressure transducers have a pressure range of ±5"H_20 and an accuracy of 0.15% of the full scale. The pressure transducer measurements are read in by a PC-based DAQ card at a sampling frequency of 50Hz where the DAQ card consisted of a 16-bit analogue to digital (ADC) converter and the software used to log the wind tunnel data for each test was LabView. To measure applied wing angle of attack, a potentiometer was connected to the turning knob of the external balance. The potentiometer is wired to the DAQ card and the angle of attack settings are recorded in LabView. The sideslip β was simply recorded from the scale on the sting bar (Fig 7.11). Digital acquisition of β settings was not necessary as dynamic tests were implemented for changes in α and not β.

In conclusion using the available instrumentation, the following parameters can be individually recorded (or calculated); p_i, P_0, P_∞, V_∞, α, β where p_i is the pressure measured from orifice i and i=1,2...,5. All wind tunnel data was filtered using a 2nd Butterworth low pass filter (implemented in LabView). To comply with the Nyquist sampling theorem, the cutt-off frequency chosen must be smaller than half the sampling frequency. With a sampling frequency of 50Hz, the cutt-off frequency was heuristically chosen to be 1.5Hz.

Fig. 7.18 Low speed wind tunnel at Leicester University laboratory

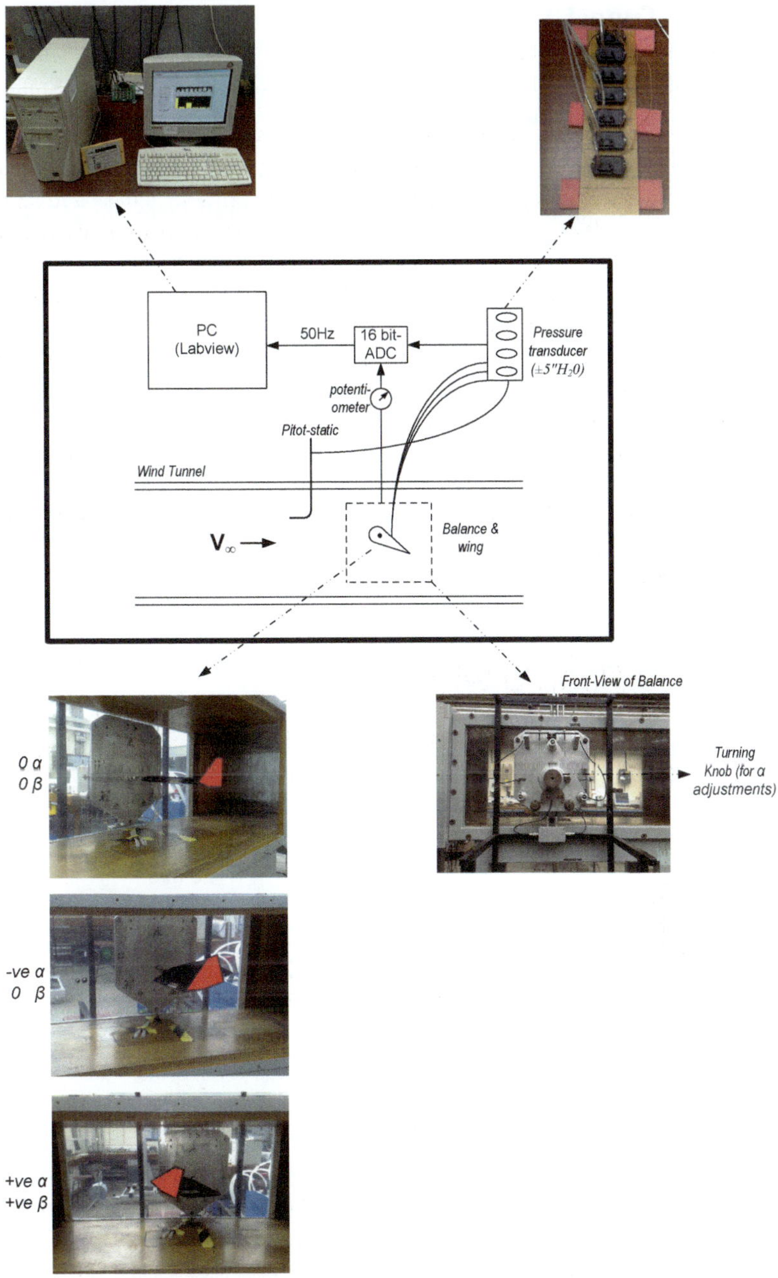

Fig. 7.19 Wind tunnel settings and instrumentation

7.6 Wind Tunnel Test Procedure

Two types of wind tunnel tests were carried out; *static* and *dynamic*. Static tests involve fixing the wind tunnel settings (i.e. V_∞, α, β settings) and recording the data *only* once the measurements have reached a steady-state value. The data can then be stored (in LabView) and analysed offline in Matlab/Simulink. The static tests carried out are outlined in Fig 7.20. Overall there are 252 separate tests with different V_∞, α, β settings. So for example, in Fig 7.20 one of the tests would consider β=0°, V_∞=15m/s and α =-9°.

The second group of tests are dynamic tests where $\dot{\alpha} \neq 0$. Static tests are useful as a first step towards analysing the NN modelling capabilities and to investigate whether or not the wing surface pressure can be related to the air data states $(P_\infty, V_\infty, \alpha)$ as predicted by the CFD simulations (section 7.2). However in real flight, static test results are of little use as the aerodynamic state of the aircraft rarely stays constant. For this reason it is important to perform dynamic tests. In our case this was considered by continuously varying the angle of attack for each test (i.e. $\dot{\alpha} \neq 0$) and fixing the airspeed and sideslip at V_∞=15m/s and β=0° respectively. The wing angle of attack is randomly varied at different $\dot{\alpha}$ and waveforms; square-wave, sine-wave and ramp-type. Note that certain time evolutions of $\dot{\alpha}$ may not be feasible in real flight, but were necessary to analyze the modelling capabilities of the NN.

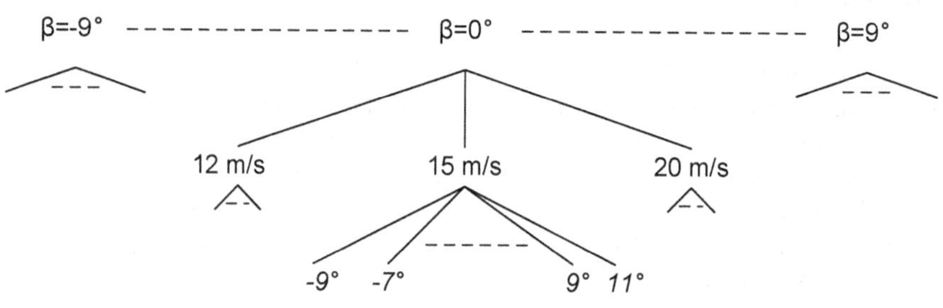

Fig. 7.20 Wind tunnel static tests. 7 different β settings (in 3deg increments), 3 different V_∞ settings, and 12 different α settings (in 2deg increments). Overall there are 252 separate static tests.

7.7 Wind Tunnel Data

As discussed in section 7.3 it is important that 1) the pressure ports (P1, P2, P3, P4, P5) give different pressure measurements and 2) these pressure measurements vary significantly with the air data states $(P_\infty, V_\infty, \alpha)$. Fig 7.21 and Table 7.3 show some of the wind tunnel data recorded during the tests:

- **Fig 7.21:** This plots the changes in the port measurements (P1, P2, P3, P4, P5) with a change in α and V_∞. From Fig 7.21 we notice:

  a) Ports give different pressure measurements which is important to avoid having redundant information
  b) Steep pressure gradients are observed at each port for a change in angle of attack setting
  c) For fixed angle of attack the pressure measurements at each port vary with the different airspeed settings.

- **Table 7.3:** This shows some of the wind tunnel data for two different sideslip settings; β=0deg and β=9deg. The observations made from Fig 7.21 are also seen in Table 7.3:

  a) For example in the first row of Table 7.3(a) we can see that the pressures measured from P1, P2, P3, P4, P5 are different
  b) The pressure measured from each port varies significantly with a change in angle of attack setting, e.g. compare the first two rows of Table 7.3(a) where for example P3=4.01 Pa for α=-9.12deg and when alpha is changed to α=-7.02deg , then P3=26.87Pa
  c) Compare the first row of Table 7.3(a) with the 13th row (i.e. when the airspeed is changed to 15 m/s), we will notice that pressure port measurements have changed significantly, e.g. P4 =-215.66 Pa for V_∞=12.73 and when airspeed is increased to V_∞=15.50, then P4=-324.31 Pa.

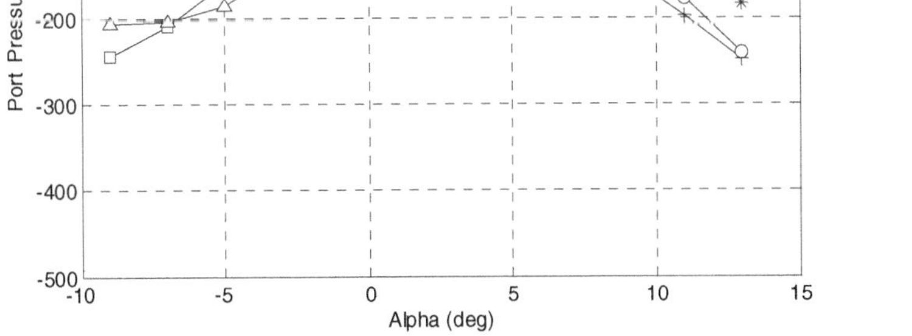

(a)

Fig. 7.21 Port pressure distribution for three different speeds and fixed β=0° (a) V_∞=12 m/s (b) V_∞=15 m/s (c) V_∞=20 m/s

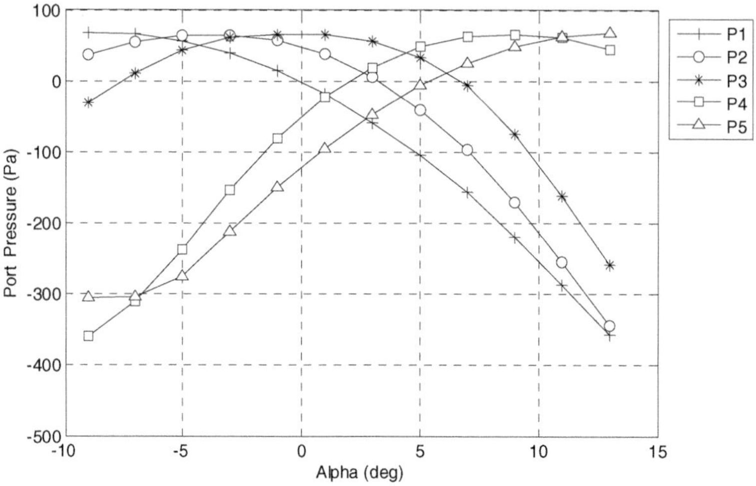

(b)

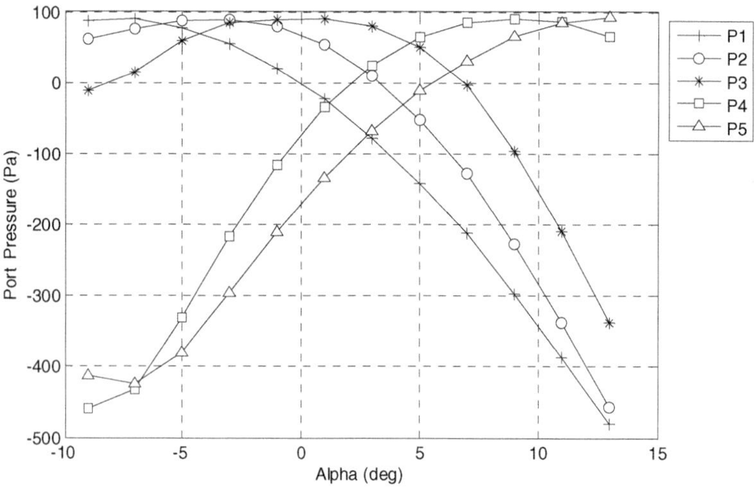

(c)

Fig. 7.21 *(continued)*

Table 7.3 Wind tunnel static test data. Note that all pressures are referenced to atmospheric pressure (101.3kPa)

(a) β=0deg

P1 (Pa)	P2(Pa)	P3(Pa)	P4(Pa)	P5(Pa)	V_∞(m/s)	α(deg)	P_∞(Pa)	P_0(Pa)
46.01	32.04	4.01	-215.66	-199.08	12.73	-9.12	-46.64	52.67
44.19	46.65	26.87	-166.03	-187.43	12.41	-7.02	-47.82	46.53
29.97	43.46	38.64	-119.69	-152.14	12.82	-4.90	-45.04	55.68
16.58	40.26	43.61	-67.17	-111.41	12.73	-2.93	-46.06	53.17
-3.39	32.43	46.26	-21.05	-72.24	12.35	-0.93	-48.20	45.14
-27.51	13.72	41.34	5.11	-40.16	12.58	1.36	-47.94	48.96
-60.72	-15.19	29.10	25.75	-10.14	12.71	3.15	-47.84	51.06
-95.50	-56.72	7.23	41.66	15.10	12.71	4.94	-47.98	50.92
-133.68	-96.94	-35.98	45.97	27.19	12.69	7.25	-47.02	51.54
-175.33	-149.28	-88.04	42.64	40.42	12.65	9.51	-47.96	50.10
-235.38	-222.57	-165.47	28.81	47.75	12.42	11.57	-47.14	47.36
-283.01	-286.07	-230.47	11.58	42.49	12.56	13.60	-47.12	49.46
62.29	46.95	-4.18	-324.31	-293.70	15.50	-9.04	-71.37	75.74
54.58	59.03	31.52	-252.55	-276.24	15.25	-7.08	-70.21	72.27
45.60	61.92	53.03	-185.86	-234.63	15.21	-4.77	-69.97	71.67
22.49	62.54	62.42	-103.36	-166.32	15.34	-2.89	-69.99	74.14
-6.58	46.60	61.29	-38.15	-103.26	15.31	-1.13	-70.23	73.24
-38.05	19.77	60.57	9.12	-59.29	15.49	1.12	-70.07	76.89
-83.32	-20.91	45.28	35.15	-22.11	15.23	3.28	-69.65	72.33
-135.96	-69.59	10.08	55.60	14.96	15.38	5.20	-70.55	74.26
-190.69	-136.18	-46.80	62.33	41.91	15.23	7.41	-69.55	72.52
-262.42	-219.20	-120.42	62.86	54.43	15.06	9.44	-69.97	68.95
-328.49	-307.18	-217.25	56.43	64.30	15.48	11.66	-70.77	75.93
-396.84	-392.77	-312.03	21.83	65.12	15.35	13.81	-69.41	74.89

Table 7.3 (*continued*)

(b) β=9deg

P1 (Pa)	P2(Pa)	P3(Pa)	P4(Pa)	P5(Pa)	V_∞(m/s)	α(deg)	P_∞(Pa)	P_0(Pa)
49.34	37.93	1.29	-225.80	-208.08	12.50	-8.85	-47.54	48.21
46.54	46.24	25.94	-172.69	-186.84	12.76	-6.88	-47.18	52.55
33.43	42.00	43.16	-116.38	-153.00	12.63	-4.94	-47.32	50.42
19.73	42.83	48.96	-66.42	-110.23	12.67	-2.95	-48.80	49.51
-1.09	35.27	46.48	-21.85	-71.61	12.67	-1.00	-48.26	50.05
-23.90	11.59	36.98	2.57	-33.01	12.61	0.81	-39.31	58.01
-57.33	-15.34	32.51	34.61	-7.34	12.50	3.08	-48.08	47.67
-95.23	-52.34	5.36	41.26	14.75	12.77	5.26	-46.88	52.92
-138.50	-99.94	-34.89	47.48	33.48	12.79	7.40	-47.64	52.49
-185.47	-156.36	-90.87	45.79	46.79	12.44	9.17	-48.44	46.32
-235.97	-220.61	-159.09	35.29	47.97	12.76	11.31	-46.90	52.75
-291.17	-289.49	-236.02	20.01	51.80	12.62	13.55	-47.62	49.94
68.87	56.75	1.50	-330.86	-302.98	15.28	-9.12	-70.21	72.85
62.09	62.67	35.62	-255.38	-284.35	15.31	-7.02	-70.43	73.04
49.07	72.72	64.33	-171.84	-228.99	15.15	-5.13	-70.67	69.91
25.13	64.25	69.06	-100.46	-168.45	15.51	-3.31	-71.65	75.63
-3.79	51.31	71.45	-36.83	-111.44	15.17	-1.04	-69.67	71.23
-39.70	23.96	61.78	7.15	-62.97	15.34	1.03	-71.23	72.82
-87.02	-20.57	47.24	45.48	-13.75	15.60	3.05	-70.03	78.99
-139.10	-75.51	11.12	63.71	20.59	15.19	4.90	-70.73	70.67
-195.88	-146.55	-48.46	68.10	41.97	15.20	7.20	-69.73	71.84
-274.12	-267.31	-222.35	-125.01	64.47	15.49	9.27	-70.25	76.62
-347.09	-314.72	-221.40	57.82	69.16	15.27	11.55	-69.41	73.32
-417.379	-424.246	-331.232	30.46	72.484	15.424	13.627	-70.49	75.22

7.8 FADS System Results

7.8.1 Static Tests

The wind tunnel data (some of which have been shown in Fig 7.21 and Table 7.3) is analysed offline in a Matlab/Simulink environment. The 252 static tests (Fig 7.20) were divided into a NN training set and a NN testing set by taking α slices, giving:

- TrD: This is the training set which includes the alpha settings of -9°, -5°, -1°, 3°, 7°, 11°.
- TeD: This is the testing set which includes the alpha settings of -7°, -3°, 1°, 5°, 9°, 13°.

Note that in each case all the β settings (-9 to 9deg) and V_∞ settings (12, 15, 20m/s) are considered. Therefore 126 static tests are used to train the NN with learning switched **on**, and the remaining 126 are used to query the trained NN with learning switched **off**. Note that the NN is trained according to the criteria defined in Chapter 4, section 4.2.

In conclusion NN training is stopped if ΔRMS< 0.1% for more than 100 consecutive epochs (for both TrD and TeD) and/or the RMS estimation error for TeD increases for more than 100 consecutive epochs. These criteria check for NN structural convergence and avoid the over-fitting phenomenon respectively (Chapter 4, section 4.2).

The NN learning rate chosen is also crucial as a high learning rate guarantees good estimations but it also degrades the global approximation capability of the network. There is no formal guideline to defining the optimum learning rate and other tuning parameters in a NN [35]. The designer must apply a heuristic-based approach when choosing the NN tuning parameters. Satisfactory performance is judged based on the estimation characteristics and execution speed of the NN. The EMRAN RBF NN tuning parameters used in our case are defined in Table 7.4.

Table 7.4 EMRAN RBF NN tuning parameters

Tuning parameter	Value
Learning rate	0.2
E1	0.2
E2	0.1
ε_{max}	0.6
ε_{min}	0.3
γ_{df}	0.997
k_{op}	1e-6

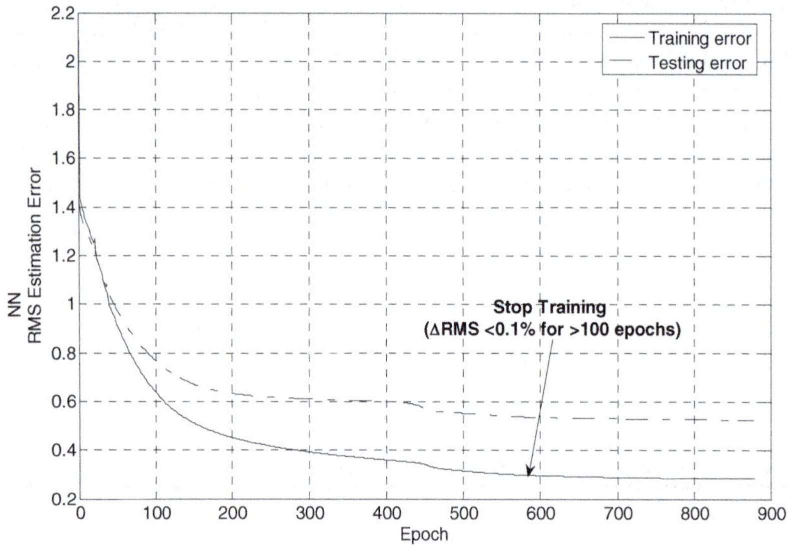

Fig. 7.22 NN training/testing RMS estimation errors

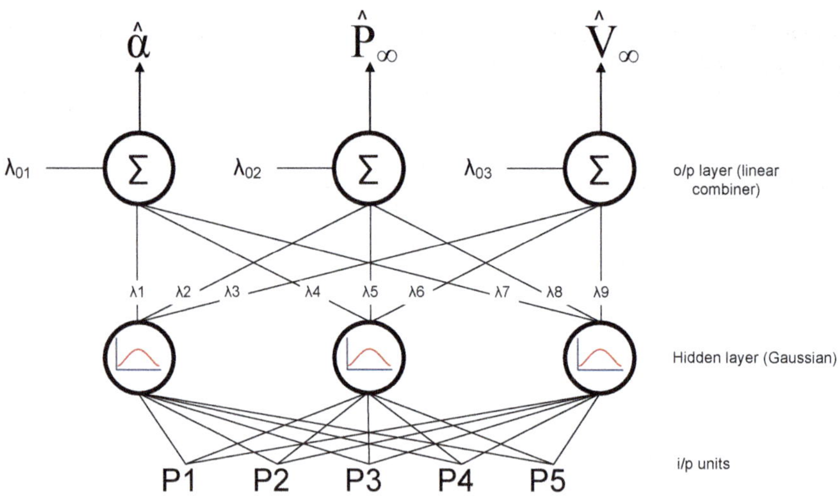

Fig. 7.23 RBF NN with 5-3-3 structure. λ's are the weights. P's represent the pressure ports in Fig 7.11.

Fig 7.22 shows the NN training stage. Note that the estimation error units are not included as it constitutes mixed NN outputs (i.e. the average of the estimation error for $\hat{P}_\infty, \hat{V}_\infty, \hat{\alpha}$). Training is stopped after 582 epochs and the NN structure frozen. The resultant NN is a fully connected 5-3-3 NN shown in Fig 7.23.

The NN testing stage involves querying the 5-3-3 NN (Fig 7.23) with the testing data set TeD. The results are broken down in Fig 7.37. Overall, the NN RMS estimation errors were; **0.44 lb/ft^2**, **0.62 m/s** and **0.51°** for $\hat{P}_\infty, \hat{V}_\infty, \hat{\alpha}$ respectively.

7.8.2 Fault Accommodation

In general FADS systems can be more robust to noisy pressure measurements in comparison to conventional air data booms [41]. This is due to the fact that multiple pressure ports are used in the FADS system and therefore averaging redundant measurements can reduce the effects of noise, while in air data booms there are only two pressure ports (total and static pressure, Chapter 3, section 3.1). Furthermore using a large matrix of pressure measurements can improve the fault tolerance capabilities of the FADS system. For example one can imagine that if 50 pressure ports, distributed over the aircraft surface, are used to compute the angle of attack then a fault in one of the ports should not significantly degrade the estimation capabilities of the FADS system.

In Chapter 2 we discussed that real systems are prone to faults (sensor, actuator faults, etc.). We also proposed the different approaches for detecting, isolating and accommodating such faults. FADS systems will also be prone to faults such as; pressure orifice blockage due to environmental conditions, pressure transducer faults and electrical wiring failures. Fault detection, isolation and accommodation (FDIA) is important if the air data are to be constantly available throughout the entire flight and especially if they are required in feedback control loops. There have been several methods cited in the literature which propose to detect faults in the pressure ports. The simplest of methods involves fault detection via physical redundancy [148]. In this case the pressure ports are divided into groups. Each group would individually estimate the same air data quantity. Their estimates can then be compared via a simple voting scheme for fault detection purposes. This technique would require a minimum of three pressure port groups to be able to implement the voting scheme. Other techniques involve statistical methods which check for the consistency of the air data estimates [43, 41, 149]. In this case the estimated air data states are re-inserted into the air data system model (see Eqs. 3.10-3.11 in Chapter 3, section 3.3) and the calculated pressure measurements are compared to the *real* pressure measurements to give the mean squared estimation error (MSE). The MSE of one set of pressure ports, and the MSE of another set of redundant pressure ports are compared, and the set with the lowest MSE is chosen in the FADS system [43].

As we can see most of the techniques for FDIA in FADS systems are based on physical redundancy. However, as discussed in Chapter 2, physical redundancy can suffer several drawbacks, such as cost, space and weight implications (which must be taken into account especially in small MAVs). As a result, new research trends have made use of model-based FDIA techniques; e.g. observers, Kalman filters, NNs etc. (see Chapter 2). The concept of model-based FDIA can also be applied in the FADS system. In this section we will compare two fault accommodation techniques: physical redundancy and NN-based approaches. Note that we are only interested in the fault accommodation stage of FDIA, i.e. we have assumed that the fault has already been detected and isolated. This is because we have already discussed and applied different FDI algorithms in Chapters 5-6. So for example we could use a NN model (an autoassociative NN, discussed later) to estimate the pressure measurements from each port (P1, P2, P3, P4, P5). The model estimates and the real pressure measurements can then be used to compute a fault residual which can be checked against a pre-defined threshold for fault detection purposes. However in this section we are only interested in the fault accommodation stage; i.e. how can we accommodate for the faulty pressure port to maintain accurate estimation performance of the NN in Fig 7.23.

To investigate this further, we artificially introduce faults in the NN testing data set TeD. The fault type considered is a *total sensor failure*. In this case the sensor stops working and outputs a constant zero. As we are so far considering only static tests, the time evolution of the fault is not of interest.

The trained 5-3-3 NN structure (Fig 7.23) is queried with TeD but this time; 1, 2 or 3 of the NN inputs are set to zero (simulating single and multiple sensor failures). Three fault accommodation techniques are then compared:

1. **No correction**: Faulty inputs are not accommodated.
2. **Next port:** In this case we simply accommodate the fault by replacing the faulty port with a neighbouring non-faulty port.
3. **AA-NN:** The faulty port is replaced with the corresponding estimate from an *autoassociative* NN (AA-NN).

The 'No correction' method represents the scenario when the fault is unaccommodated. The 'Next port' method represents the *physical redundancy* technique for fault accommodation purposes. So for example in Fig 7.11 if P1 is faulty, then measurements from P2 can replace it. The NN in Fig 7.23 will therefore have pressures ports; P2, P2, P3, P4 and P5 as inputs. If P1, P3 and P5 are faulty then the NN inputs would be P2, P2, P4, P4 and P4. The 'AA-NN' method for fault accommodation represents the *model-based* approach. An AA-NN is one which simply reproduces its inputs at its output [97]. In other words the AA-NN would be trained to receive measurements from P1, P2, P3, P4, P5 and produce the corresponding estimates $\widehat{P}1, \widehat{P}2, \widehat{P}3, \widehat{P}4, \widehat{P}5$. Fault accommodation via AA-NN is then implemented by replacing the faulty port with the corresponding estimate from the AA-NN. This method of fault accommodation was originally proposed in [36] and shown to improve the FADS system robustness to sensor faults. In our case, the AA-NN designed is a fully connected 5:10:5 EMRAN RBF NN.

Fig 7.24 shows the results for the fault accommodation tests. In general we notice that if the fault is not accommodated (i.e. the 'No Correction' option), the NN RMS estimation errors increase significantly especially in the presence of multiple faults (with alpha RMS error reaching 14° for three faults, Fig 7.24 (c)).

However in the 'Next Port' option, estimation errors are reduced significantly. In our case using the 'Next port' option resulted in an average **50.39%** decrease in the NN RMS estimation errors. On the other hand using the 'AA-NN' option resulted in a larger reduction of **69.6%**. The AA-NN greatly improves the robustness of the FADS system to faults.

A drawback of using the AA-NN is the further memory usage required onboard the air vehicle. On the other hand, the 'Next Port' option does not suffer from this and is simpler to implement which is why in many cases the FADS system is pre-designed with a slight degree of pressure port redundancy. In our case (Fig 7.11) we can see that the pairs P1, P2 and P4, P5 could be considered as redundant port locations as 1) They are close to each other 2) They have similar pressure distributions as seen in Fig 7.21.

In conclusion, including redundant pressure ports in the FADS system improves the robustness of the overall system to faults. Furthermore it can mitigate the effects of system noise. However in applications where space onboard the vehicle is limited or instrumentation costs need to be low, an AA-NN can be designed to reproduce the faulty (or noise-corrupted) pressure measurements.

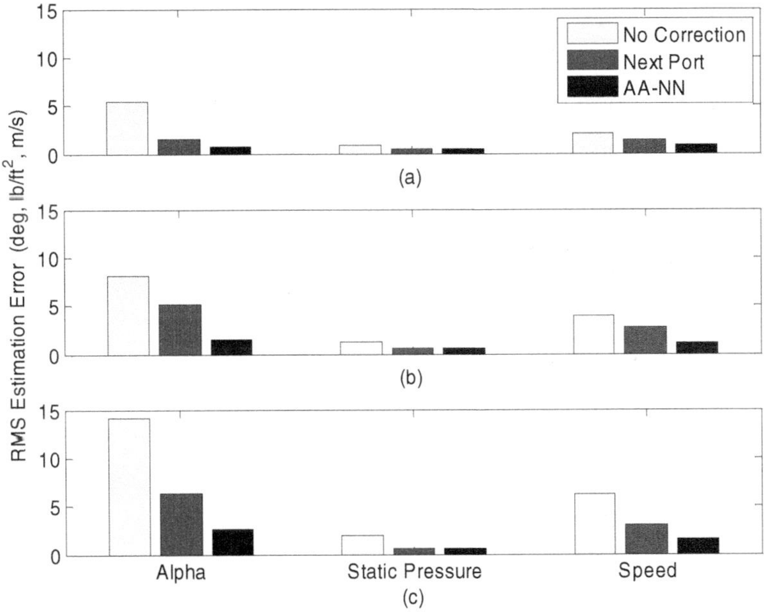

Fig. 7.24 NN RMS estimation errors in the presence of total sensor failure. (a) P1 faulty, (b) P1, P5 faulty, (c) P1, P3, P5 faulty.

7.8.3 CFD vs. Wind Tunnel

In section 7.2 we concluded that the pressure ports P1, P2, P3, P4, P5 were insensitive to changes in the sideslip β (Fig 7.16 and Fig 7.17). To validate the CFD results we can also plot Cp vs β from the wind tunnel data. Such a plot is shown in Fig 7.25(a) where the CFD predictions and the wind tunnel data are both

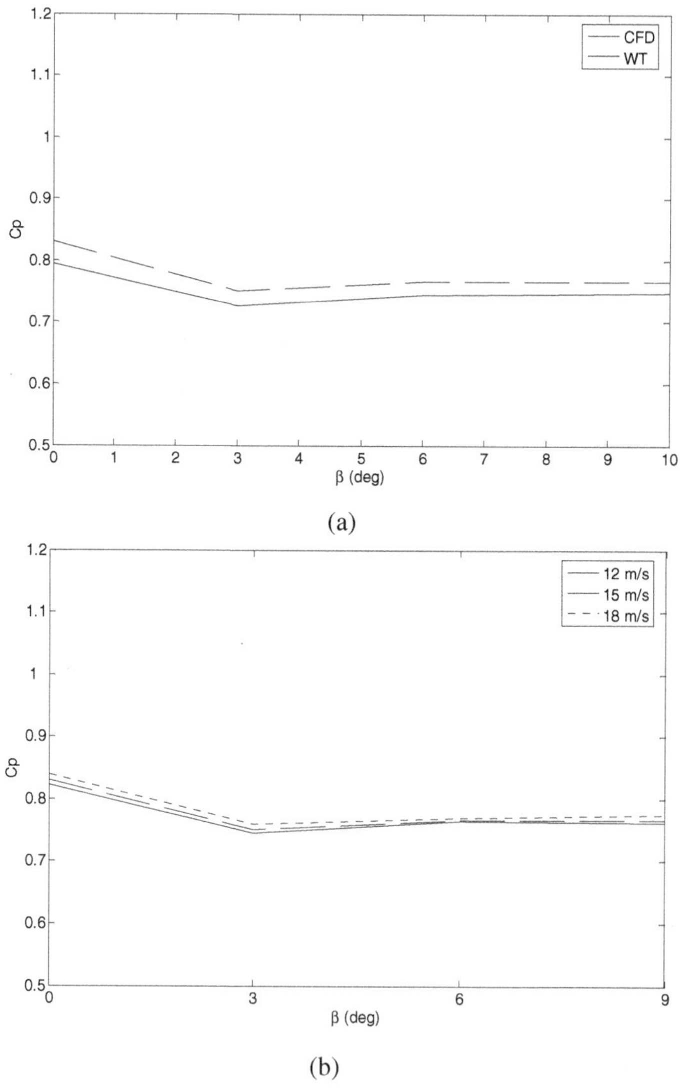

Fig. 7.25 (a) CFD and wind tunnel data for pressure port P3 ($\boldsymbol{V}_\infty$ = 15m/s, $\boldsymbol{\alpha}$=0deg) (b) Wind tunnel data when changing the airspeed $\boldsymbol{V}_\infty$ = 12, 15, 18 m/s

plotted for pressure measured from P3. There are two important observations to make from Fig 7.25(a):

1. CFD and wind tunnel data are almost similar: There will be some differences due to the assumptions considered in the CFD simulations (e.g. inviscid flow and the dimensional assumptions such as constant wing thickness, see section 7.4).
2. Both the CFD and wind tunnel data show that the variation of pressure with sideslip is insignificant.

Fig 7.25(b) also shows the wind tunnel data for a change in β for 3 airspeed settings. From both Fig 7.25(a) and Fig 7.25(b) we can conclude (as we had earlier predicted from the 3D-CFD results, section 7.4) that the pressure ports located as in Fig 7.11 are insensitive to changes in sideslip. Therefore it can be difficult to define a NN model which relates the surface pressure to the sideslip, as this relationship is almost constant (see Fig 7.25).

7.8.4 Dynamic Tests

The 5-3-3 NN structure concluded in Fig 7.23 is used in the dynamic tests. At a fixed β=0° and V_∞= 15 m/s, the wing angle of attack is varied continuously (i.e. $\dot{\alpha} \neq 0$) and the NN estimate $\hat{\alpha}$ recorded. An overall alpha RMS estimation error of **0.58°** was achieved.

Fig 7.26 and 7.27 show the NN estimation characteristics for two of the dynamic tests carried out. Let us refer to them as *dynamic test 1* (DT1) and *dynamic test 2* (DT2) respectively. Let us now investigate the estimation characteristics of DT1 in detail (Fig 7.26). It can be seen that at certain time frames the NN estimations are significantly poor. For example at around 120 seconds, where α>15°, the NN underestimates α by almost 10 deg. At first it may seem that this is simply due to random estimation error patterns, i.e. the NN performance is generally accurate except for some random time frames. However if we observe the estimations from DT2 (Fig 7.27) we notice that at approximately 75 seconds, where again α>15°, the NN underestimates α by almost 10deg. This shows that the NN performance is in fact poor for specific α settings, such as α>15°.

One explanation for this observation is related to the NN domain of validity [35]. The RBF-NN is essentially a multidimensional interpolator. The accuracy of the surface fit to 'new' data (i.e. its ability to generalise) is highly dependent on the location of this data in the input space, i.e. the location of that point relative to the domain of validity of the NN where the *domain of validity* defines the boundaries of the data set used to train the NN [150]. A general rule is that when the NN is queried with data which lies outside these boundaries (i.e. outside the domain of validity), the NN estimations would be poor or in other words the extrapolation properties of the NN are poorer than its interpolation properties. This can be better explained by re-iterating the RBF NN structure (Chapter 4). The fundamental principle of the RBF NN is that the output of each RBF found in the hidden layer (i.e. the output of each hidden neuron) is proportional to the *distance* between the input vector and the *centre* of that hidden neuron. So for example in our case we chose a Gaussian- RBF. Therefore data which lies far from the centre of the Gaussian function will output a close-to-zero value. These centres are tuned based on the training data set, TrD. Therefore testing data which is far away from the domain of TrD will also lie far from the centres of the Gaussian functions and so the NN output will always be close to zero. In other words the NN estimations will be poor for testing data lying outside the domain of TrD, i.e. outside the domain of validity.

For this reason it is important that the NN is trained with as much flight data as possible, and more importantly with data which covers the entire flight data range. This reduces the likelihood of the testing data lying outside the NN domain of validity. In our case, the method of dividing the training (TrD) and testing (TeD) data sets was chosen for ease of presentation. However, a more robust approach is needed to define TrD, so that the domain of validity encompasses all possible input data patterns. This is particularly important in our case as the NN structure is frozen after the offline training stage and will not be further trained during actual flight.

In conclusion, a more robust approach to training the NN must include the following:

1. Collecting all the possible flight data (this would depend on previous flight experience)
2. Defining the domain of validity of the collected flight data
3. Including the outermost values of the domain of validity to train the NN.

A good survey paper for the different methods to defining the domain of validity can be found in [151]. The simplest approach is by defining the minimum and maximum of each NN input parameter (in our case that would be each pressure port). Testing data which lies outside these limits is said to be outside the domain of validity. Despite its simplicity, unfortunately this method overestimates the domain of validity. More complex methods involve defining the convex hull (CH) of the training data set. The vertices of this convex hull can then be used to determine the exteriority of a new input pattern. Patterns which lie outside the CH have an exteriority greater than zero, and ones which are in the CH have an exteriority equal to zero. A convex hull is a multidimensional surface which tightly encompasses a set of multidimensional data [35]. There are several techniques to defining the convex hull and its vertices. They are generally based on iterative mathematical algorithms which continuously check the exteriority of a new input pattern. If this exteriority is greater than zero then the input pattern is included in the convex hull and so on. See for example Chapter 4.4 in [35] for a code which calculates the vertices of the convex hull. Another method of calculating the domain of validity is by defining a sphere which encompasses all the input data patterns. Data patterns which lie outside the sphere are considered to be outside the domain of validity [150]. As in the min/max method, this approach is simple to implement as we would only need to know the radius of the sphere to define whether the input pattern lies outside the domain of validity. However, similar to the min/max method, this technique can grossly overestimate the domain of validity.

Regardless of the technique used to defining the domain of validity, the important thing to note is that once it has been defined, the vertices (i.e. the outermost values) of the domain of validity must be included in the NN training set as this greatly reduces the need for NN extrapolation. As wind tunnel tests generally investigate only a group of flight test conditions, it is difficult to define the domain of validity based on only the wind tunnel data.

To demonstrate the importance of carefully selecting the NN training set TrD, we will reconsider the NN estimations for DT1 in Fig 7.26. For simplicity purposes, let us define the domain of validity using the min/max approach described earlier. We will assign pressure from TrD and DT1 the subscripts '_TrD' and '_DT1' respectively. The following code (shown for P1 only) can be used to indicate (graphically), whether a pressure input is outside the domain of validity defined by P1_TrD (i.e. defined by the min/max range of the pressure from P1 in the training set TrD):

```
P1_TrD is the pressure from port 1 in the training set TrD
P1_DT1 is the pressure from port 1 in the dynamic test DT1

IF P1_DT1 is less than the minimum value in P1_TrD or P1_DT1 is greater than the
   maximum value in P1_TrD THEN
   This indicates that P1_DT1 is outside the domain of validity.
   Therefore plot the number 25.
ENDIF
```

The code states that if pressure from P1_DT1 exceeds the bounds defined by P1_TrD then plot a value of 25. The plot value is not meaningful, it is chosen simply for ease of representation. Similarly the code above can be implemented for the remaining pressure ports P2, P3, P4 and P5 but with plot values of **20, 15, 10 and 5** respectively.

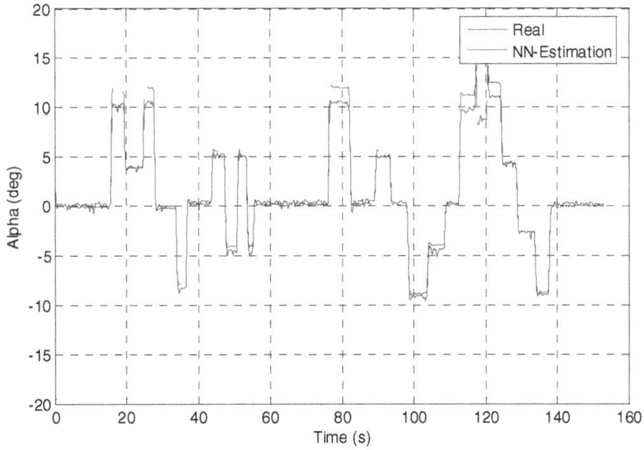

Fig. 7.26 RBF NN estimation ($\hat{\alpha}$) for DT1

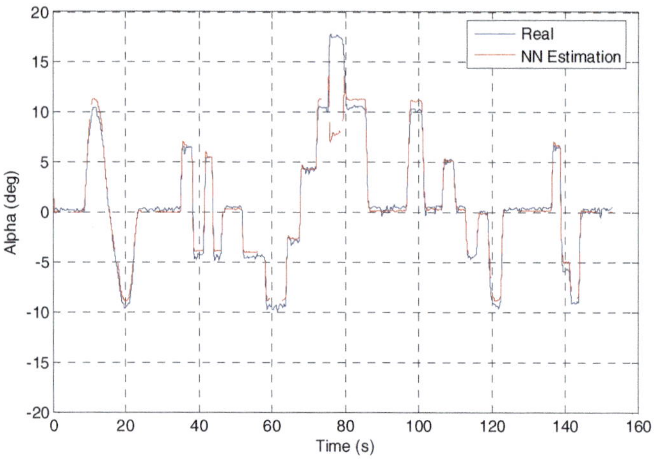

Fig. 7.27 RBF NN estimation ($\hat{\alpha}$) for DT2

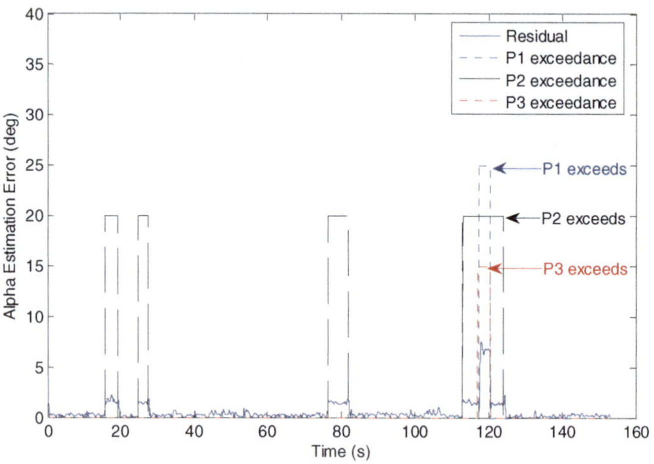

Fig. 7.28 Domain of validity test for P1, P2, P3. If P1, P2, P3 lie outside the domain of validity, a value of 25, 20, 15 is plotted respectively, otherwise a value of zero is plotted. NN residual also shown (solid blue line~'1)

Fig 7.28 shows the results from this task. We have also included the residual from DT1 which was calculated simply as the *difference* between the NN estimates $\hat{\alpha}$, and the measured (real) value α. Note also that the plots, in Fig 7.28, for P4 and P5 cannot be seen as they do not exceed any limits.

From Fig 7.28 we notice how the residual significantly increases (i.e. NN estimations are poor) only when the pressure input patterns lie outside the NN domain of validity. So for example the residual is highest at 120 seconds when pressure from ports P1, P2 and P3 exceed the min/max limits, i.e. they lie outside the domain of validity.

The simple domain of validity test presented here has confirmed the importance of appropriately selecting the NN training set. Different approaches exist for doing so and it is up to the designer to choose a suitable method. It is perhaps also important to note that training data is best obtained from real flight tests as the FADS system will eventually be used during real flight. BBSR Ltd has already purchased a suitable air data boom for the MAV which can be mounted on the MAV and used to collect real flight data. This data can then be used to train the NN.

7.8.5 FADS System via LUTs

Introduction

The RBF NN designed for the FADS system is a static model i.e. the NN structure is frozen when used onboard the air-vehicle. On the other hand the NN models used in the SFDIA applications (Chapters 5-6) were dynamic, i.e. the NN structure was continuously updated via online training. Therefore as we are no longer benefiting from the adaptive capabilities of the NN we must also consider alternative static input-output mapping techniques for the FADS system model. The most popular being the lookup table (LUT). LUTs are based on storing (user-defined) input/output *training* patterns in memory. If the *testing* data patterns are exactly equal to the training patterns then the LUT simply recalls the corresponding pattern from memory. Therefore in this case, LUTs would most certainly outperform an RBF NN. However the two methods must be compared in terms of their generalisation capabilities, i.e. how they would perform when exposed to 'new' data.

A common misconception is that LUTs must visually look like a table, i.e. they have structures similar to a table. However in some cases there may be table elements which are not available or do not physically exist. So for example, the thrust from an engine can be defined by a 2D lookup table with Mach number and altitude as the independent variables. In this case, there exists a unique engine thrust value for every Mach-Altitude combination. Now consider our FADS system application. Let us assume that the row vector is the measurements from P2, the column vector is measurements from P1 and the output is the angle of attack (Fig 7.29). In Fig 7.29, a cross implies that an angle of attack value does not exist for the specified P1-P2 combination. This could be because the wind tunnel tests did not consider these angle of attack settings, or simply that the pressure combination cannot physically exist in the FADS system.

The data used to build the LUT will be referred to as *training* data and the data used to query the LUT will be referred to as *testing* data (to maintain consistency with the NN terminology).

The LUT elements (Fig 7.29) can also be plotted in a xyz Cartesian coordinate (Fig 7.30). The *generalisation* capability of the LUT is judged based on its performance (i.e. output accuracy) when the testing data does not coincide with the training data e.g. if P1=95.1 and P2=45.1 in the LUT (Fig 7.29). In this case, there is no pre-defined value for the output. To tackle this problem, LUTs are generally interpolated between the data points. For example in Fig 7.30, a 2D surface can be fitted so that there is an output for all P1-P2 combinations lying between the training data set, see e.g. Fig 7.30(b)-(c). This approach is referred to as surface fitting (or curve fitting in the 1D case) and belongs to the field of computational geometry [152, 153]. In our case, as we are considering LUTs, an

Column vector (↔)

Row vector (↕)

P2 \ P1	90	95	100
40	1deg	×	3deg
45	×	2deg	×
50	×	×	4deg

Fig. 7.29 2D LUT example for angle of attack

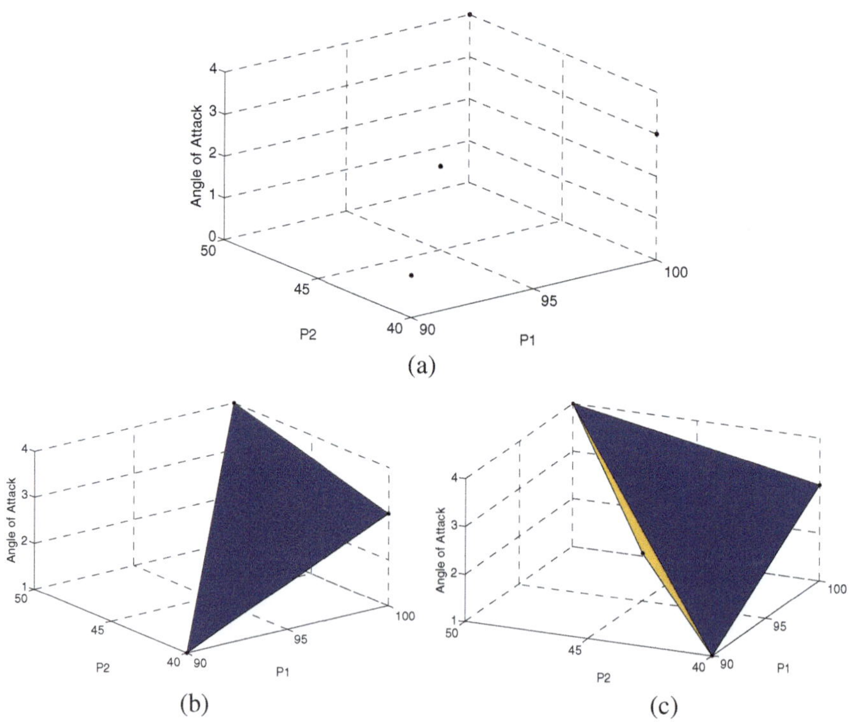

Fig. 7.30 Surface fit for 2D- LUT in Fig 7.29 (a) Data points are shown, (b) Data points triangulated to generate a surface, (c) Rotating Fig 7.30(b)

important criterion is that the surface fit **must** pass through **all** the data points (otherwise known as *strict* interpolation [97]). So for example in Fig 7.30, the surface fit must pass through all the data patterns defined in Fig 7.29 and shown in Fig 7.30(a).

The following terminology is used throughout this section:

- Data point/Input: A specific LUT input pattern (e.g. P1=90, P2=40 in Fig 7.29). Also the x-y coordinate of a surface fit such as Fig 7.30. Inputs to the LUTs are the independent variables.
- Output: This is the scalar output of the LUT (e.g. for input P1=90, P2=40 the output is 1deg in Fig 7.29). Also the z coordinate of a surface fit (e.g. angle of attack in Fig 7.30). Outputs of a LUT are the dependent variables.

Application

LUTs can be multidimensional, i.e. they can have more than one independent variable. For example the LUT in Fig 7.29 is 2D where there are two independent variables (P1 and P2) and one dependant variable (angle of attack). It is generally uncommon to find LUTs with dimensions higher than 3D. This is because they can be computationally demanding and require large amounts of memory onboard the aircraft. Instead we can divide one LUT into several low-dimension LUTs (each dedicated to estimating only one air data variable).

The FADS system consists of five independent variables (P1, P2, P3, P4, P5) and three dependent variables (angle of attack, airspeed, freestream static pressure). To simplify the problem we will design three separate 2D-LUTs (one for each dependent variable). The design is summarised in Fig 7.31.

To justify the choice of the pressure ports in Fig 7.31, let us refer back to Fig 7.11. P3 is the middle pressure port and is where the total pressure can be measured. As total pressure is a function of airspeed and static pressure (via Eq. (7.5)), port P3 will also be sensitive to airspeed and static pressure. For this reason P3 is included in LUT 2 and LUT 3. From the CFD simulations (section 7.2, see e.g. Fig 7.8), we concluded that most of the steep pressure gradients vs. angle of attack occur close to the wing leading edge, i.e. at $x/c<0.01$. The ports closest to the wing leading edge (other than P3 as it has already been used in LUT 2 and LUT 3) are P2 and P4, and are therefore included in LUT 1.

Popular univariate interpolation methods (i.e. 1D problems where there is only one independent variable) include linear interpolation (where a straight line is fitted through all the data points), polynomial interpolation (fitting a polynomial curve of specified order), linear splines (multiple straight lines are fitted to the data where each line connects 2 adjacent data points), cubic splines (3rd order polynomials connecting every 2 adjacent data points), linear regression (a curve is fitted, such as a polynomial, to the data in such a way that minimises the squared estimation error). Collectively these methods are referred to as curve fitting approaches [152]. For a wider introduction to such methods and multidimensional fitting approaches, the reader is referred to [152-155].

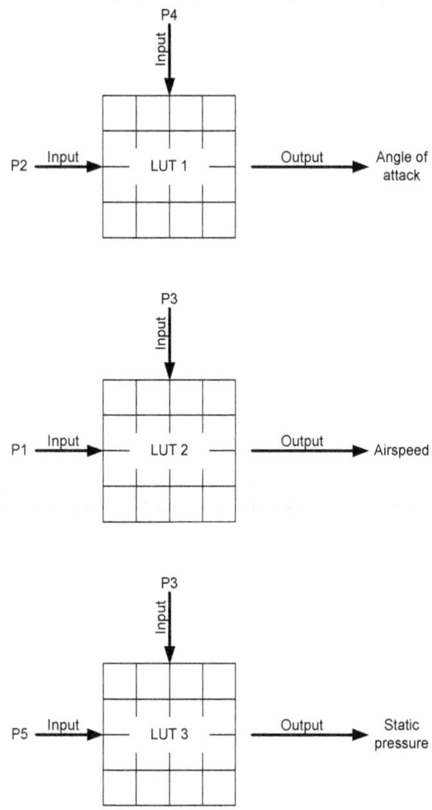

Fig. 7.31 Three LUTs used in the FADS system

In our case there are two independent variables for each LUT (Fig 7.31) and so we are considering 2D surface fitting problems such as the example in Fig 7.30(b)-(c). Surface fitting can be a challenging task as experimental data is often randomly scattered on a plane and therefore it is not clear how they should be connected, i.e. in what pattern. One popular technique is via triangulation [152]. Triangulation is the process of generating multiple triangles where each triangle has vertices coinciding with three data points. The benefits of this and how the generated surface can be used to generalise will be discussed later. More formally, the triangulation technique to be used is known as Delaunay triangulation and is in fact the 2D data gridding technique used in most Matlab functions.

Delaunay triangulation is based on a Dirichlet tessellation of the plane so that the generated triangles are as close to equilateral triangles as possible [156]. One reason for this is that equilateral triangles have the property that all three sides are equal in length. Therefore the three data points on the vertices of the triangle would be at the same distance from each other. This localises the triangulation to data points which are close to each other, i.e. to every three data points which are in close proximity. Dirichlet tessellation first divides a plane into tiles and then triangulates between data points which share the same tile 'edge'. Fig 7.32 shows such a tessellation and the resulting Delaunay triangulation. Note that in Fig 7.32 the *convex hull* of the 8 data points would be the outer-boundary $N_1N_2N_3N_4N_5N_6$. For a more formal, mathematical introduction to Dirichlet tessellation and Delaunay triangulation the reader is referred to [157, 158].

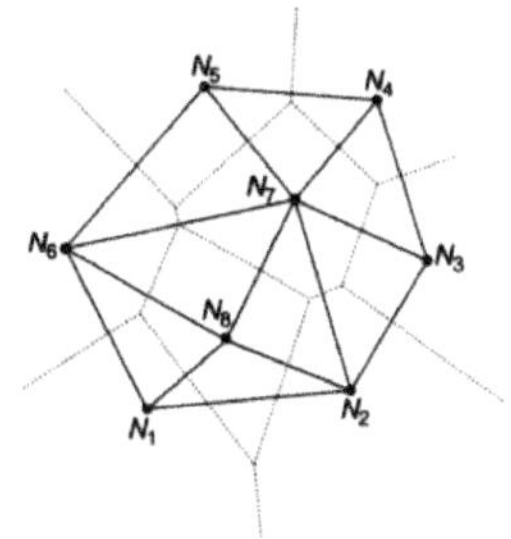

Fig. 7.32 Dirichlet tessellation (dotted lines) and resulting Delaunay triangulation (solid lines) for 8 data points N_1, N_2....., N_8 [156].

One of the motivations for triangulating a surface (as in Fig 7.32) is that each triangle can be described as a linear equation [153]:

$$z = a_0 + a_1x + a_2y \tag{7.9}$$

where z is the output (i.e. the z-axis of a surface fit, output of the LUT), x and y are the inputs (i.e. the x-y coordinates of the surface fit, inputs to the LUT). Note that in for example Fig 7.32, each data point (N_1, N_2,..., N_8) would have a corresponding x-y coordinate. Therefore for any three data points, the coefficients in (7.9) can be derived as long as the three data points are not collinear [152]. The resulting linear equation describes a triangle with vertices at these three data points. So for example in Fig 7.32, the x-y coordinates of the data points N_5, N_7, N_4 can be used to derive the coefficients a_0, a_1, a_2 in (7.9) so that the resulting triangle connects these three data points. The interpolated (triangular) surface $N_5N_7N_4$ is then used to calculate the output (i.e. z coordinate) of any 'new' data point which lies on the edge (or inside) this triangle. The latter is the process of 'generalisation' and one can imagine that the denser the triangulation (i.e. the surface is divided into many triangles) the more localised the surface patterns are, and therefore the better the generalisation capabilities of the surface fit.

An example of a triangulated surface fit is shown in Fig 7.31; the x and y coordinates would be the input to a 2D LUT and the z-coordinate is the output of the LUT. Note that if we are to refer to the 2D LUTs in Fig 7.31 as surface fits, an important criterion is that the fitted surface passes through all the data points.

The surface fit discussed so far interpolates through all the data points so that data lying inside or on the edge of the convex hull is approximated by the linear triangle equation (7.9). The problem occurs when the data lies outside the convex hull (e.g. outside $N_1N_2N_3N_4N_5N_6$ in Fig 7.32). The generalisation performance of a surface fit to data points lying outside the convex hull is then defined by its *extrapolation* properties. Earlier in section 7.8.3 we concluded that the NN performance is poorer when data lies outside the NN domain of validity (e.g. outside its convex hull), i.e. extrapolation was poorer than interpolation. The same is true for the surface fit such as Fig 7.33. The reason for this is that there are no local data patterns outside the convex hull from which we can triangulate the data. Therefore we must apply different techniques to defining the output when the input data points lie outside the convex hull. Ideally however, as discussed in section 7.8.3, we would want the training set to encompass as much of the flight data range as possible so that most of the testing data lies inside the triangulated convex hull. This avoids the need for extrapolation.

Part of the extrapolation process is to search for the closest training data points. If these data points are significantly far from the testing data point, then we would expect poor extrapolation performance. So for example in Fig 7.33, if we query the surface fit with P1=1000 Pa and P2=1000 Pa we can expect the estimated angle of attack to be inaccurate, as the P1 and P2 have a data range of only 0-15 Pa. On the other hand if the query data point is P1=15.01 Pa and P2=15.01 Pa, then this data point lies very close to the limits (i.e. convex hull) of the surface fit in Fig 7.33.

The same methods used in interpolation can also be applied to the extrapolation process. So for example once we have found the closest training data points to the testing data point, we can then fit a polynomial equation through these training data points and use it to approximate the testing data in close proximity.

In conclusion the method used to generate an interpolated surface fit for the training data points is as follows:

1. Triangulate the data points via Dirichlet tessellation so that each triangle connects 3 three data points (to satisfy the LUT criterion)
2. For each triangle use the xyz coordinate of the 3 data points to derive a linear equation as in (7.9)
3. Store the set of linear equations.

Once the interpolated surface is generated, the testing data is used to query the surface fit as follows (regardless of whether the data point lies inside or outside the convex hull):

1. Define the testing data point.
2. Find the closest triangle and its associated equation as in (7.9).
3. Use this linear equation to define the output of the testing data point

Therefore with these conditions we can define the following outcome:

1. If the testing data point is exactly equal to one of the training data points then it will share the same output value, i.e. the estimation error is zero. This satisfies the LUT criterion.
2. If the testing data point lies inside, on the edge, or outside the convex hull then the output value will depend on the closest data pattern, i.e. the closest triangle. Therefore both the interpolation and extrapolation methods are based on searching for the closest triangle.

Prior to analysing the results it must be noted that the methods chosen here are not necessarily superior to all other interpolation/extrapolation techniques. For example we may fit only one polynomial equation to approximate all the data patterns. This will be much simpler and quicker than triangulating the surface and solving (7.9) for every local triangle. In general one cannot predict the best methods by simply observing the data patterns. Different techniques must be first applied and based on specific merits (such as time consumption, memory usage, estimation accuracies, extrapolation properties etc.) we can only then define the most suitable method.

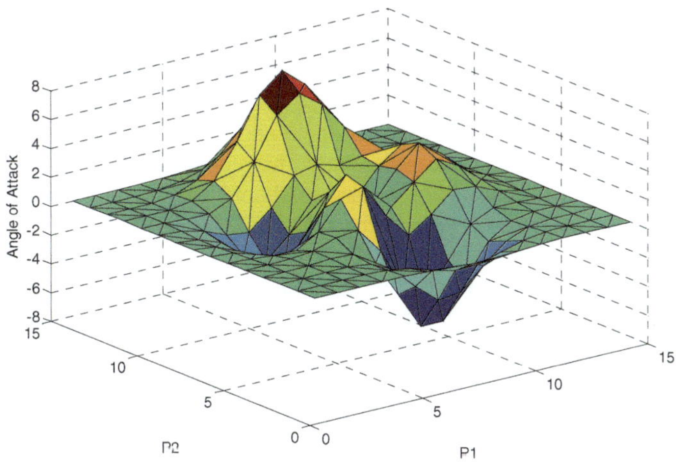

Fig. 7.33 Surface fit example using Delaunay triangulation

Results

The results for the three LUTs (Fig 7.31) are shown in Figs 7.34-7.36, Fig 7.38 and Table 7.5. A fault accommodation test was also implemented to investigate the fault tolerance capabilities of the LUTs. The tests were similar to the ones carried out in section 7.8.2 but the method of fault accommodation only involved the 'Next port' option. A careful analysis of the results reveals the following:

- **Fig 7.34:** This shows the surface fit for all three LUTs. The outer-boundary of each plot is the convex hull and testing data lying outside the convex hull are shown as black 'dots'. On average 20% of all testing data lied outside the convex hull or in other words, 20% of the data had to be extrapolated. The triangulation of the surface can be clearly seen, however we notice that the triangles are quite large. This reduces the accuracy of the surface fit as it is desirable to build a dense surface fit in order to increase the sensitivity to local patterns (see for example Fig 7.33).
- **Fig 7.35:** This summarises the overall estimation error of the LUT. In comparison to the LUT, the NN (except for static pressure estimations) outperforms the LUT. One reason for this is that the surface fit generated by

triangulation does not produce a dense surface. This could be improved by increasing the amount of training data.

- **Table 7.5:** A fault accommodation test was implemented and the LUT and NN results tabulated. The table shows the percentage increase in estimation error when a pressure port is replaced with a neighbouring port. We notice that in some cases the LUT has a 0% change in estimation error. This is because the faulty pressure port is not used in the input of the LUT, while the NN on the other hand uses all 5 pressure ports. For this reason the LUT has shown to be more robust than the NN to faults (as shown by a generally smaller mean %).
- **Fig 7.36:** This shows the average execution time for one sample of data (1.6 GHz Pentium processor used). The LUT tends to be faster than the NN. There are two possible reasons for this. Firstly, if the testing data is equal to the training data, the LUT simply recalls the data pattern from memory, i.e. no computations are required. If the testing data is not equal to the training data the LUT searches for the closest triangle and its linear equation (7.9). It then substitutes the testing data value into this equation to compute the output value. Equation 7.9 has only 4 basic mathematical operators (i.e. 2 additions and 2 multiplications). On the other hand the 5-3-3 NN involves much more mathematical computations (see Eq. 4.10 in Chapter 4) and therefore takes longer to execute. In general however both LUT and NN had a mean execution time per data sample lower than the flight data sampling time (0.02s).

In general the NN was found to outperform the LUT in terms of estimation accuracy. The main drawback of the LUT is its extrapolation properties which accounted for 20% of the testing data. In section 7.8.3 we also suggested that the 5-3-3 NN had poor extrapolation performance. The reason for this is that the RBF-NN outputs are based on the distance of the input data to the local centres of the RBFs. Therefore if the testing data are far away from any centres then the activation of each RBF would be small and the output would be close to zero. To avoid data extrapolation it is important that the training data set, used to build the LUT (or train the NN), encompasses all possible data patterns. This reduces the possibility of the testing data to lie outside the convex hulls. The size of the training set is also important. In our case, only 126 different test conditions were used to build the NN and LUT structures. In general, especially for NN applications, it is not uncommon to use thousands of different training patterns e.g. in our SFDA application (Chapter 5, section 5.8.1) we trained (offline) the NN with 225s of flight data which is equivalent to 11250 training samples (for a sampling time of 0.02s).

It was also found that the LUT has faster execution times than the NN. This is because the mathematical operations performed per data sample are much less for the LUT in comparison to the NN.

Therefore at first we may conclude that the LUT outperforms the NN according to several criteria; execution time, fault tolerance and simplicity. A logical question therefore is what can we benefit from using the 5-3-3 NN in the FADS system. Firstly, despite pointing out earlier that air data booms are not to be used in our MAV (due to weight and cost restrictions), in many other manned air-vehicles (or even large UAVs), the FADS system will operate in parallel (i.e. as a back-up) to the air data boom. This gives the option of NN online training as we now have reference air data from the boom. Although look-up tables can also be designed to be adaptive, it is the online training capability of NNs which make them extremely popular. An additional reason is that LUTs are at most 3D while it is not uncommon to find NNs using more than 50 inputs and can therefore be more robust to input faults and measurement noise in comparison to the LUT. Finally while a NN can be implemented in a few lines of code, the LUT demands much higher memory usage especially if the training data set is large.

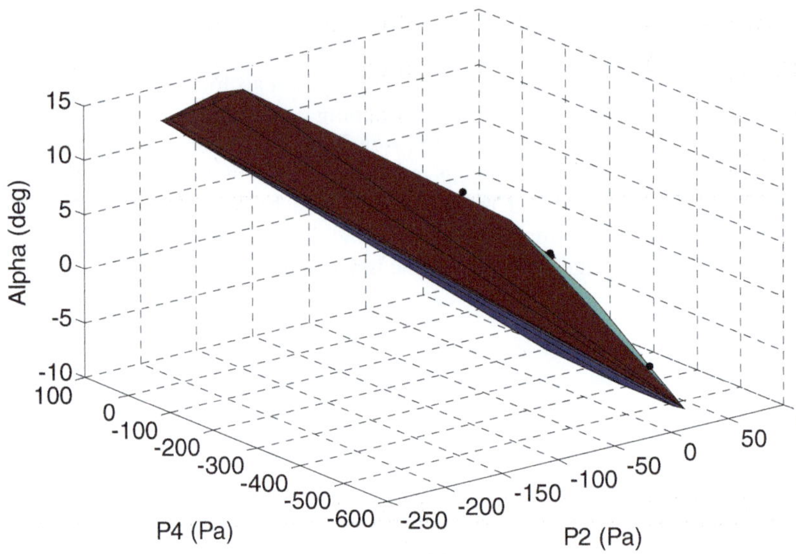

(a)

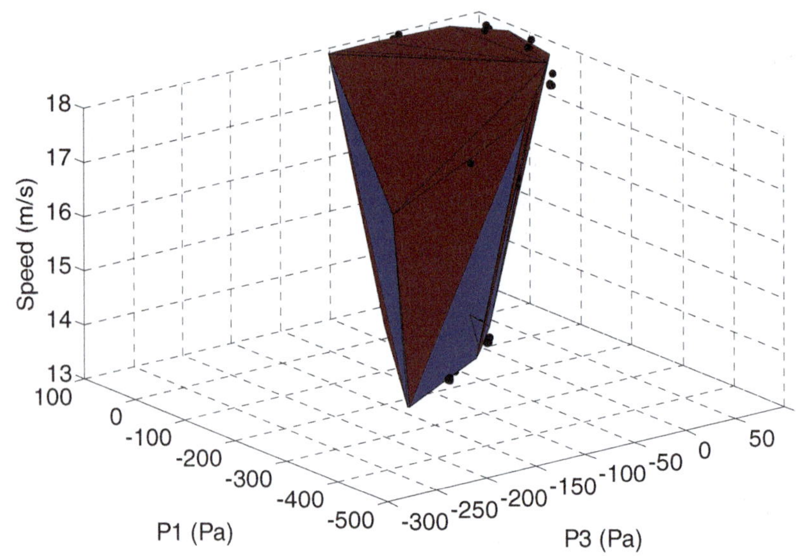

(b)

Fig. 7.34 Surface fits for (a) LUT 1 (b) LUT 2 (c) LUT 3, (black dots are testing data lying outside convex hull)

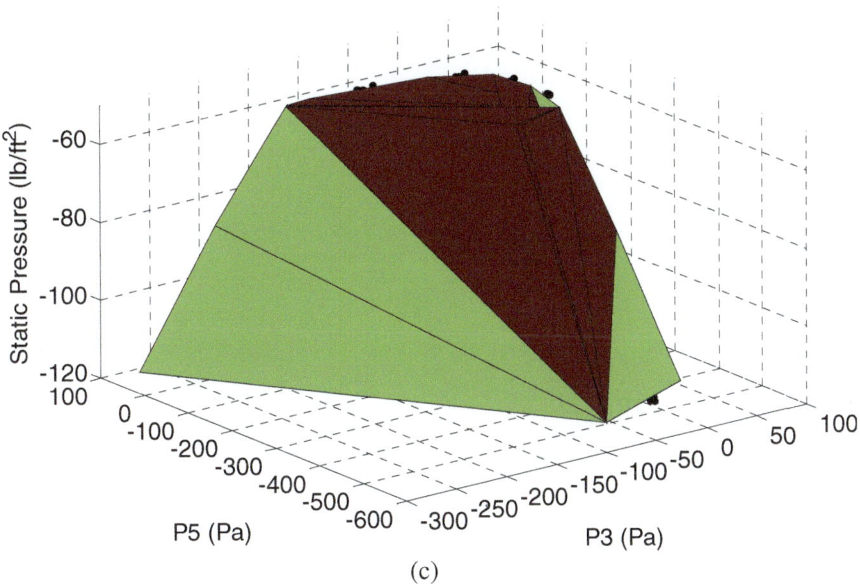

(c)

Fig. 7.34 (*continued*)

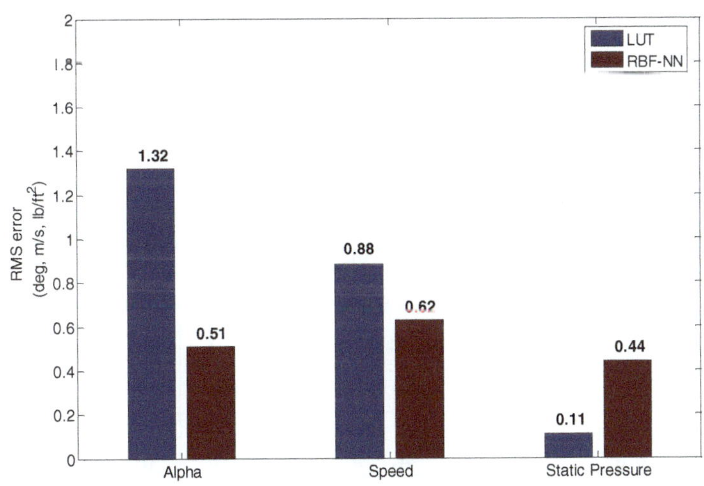

Fig. 7.35 LUT and NN average estimation error for the test data set, TeD

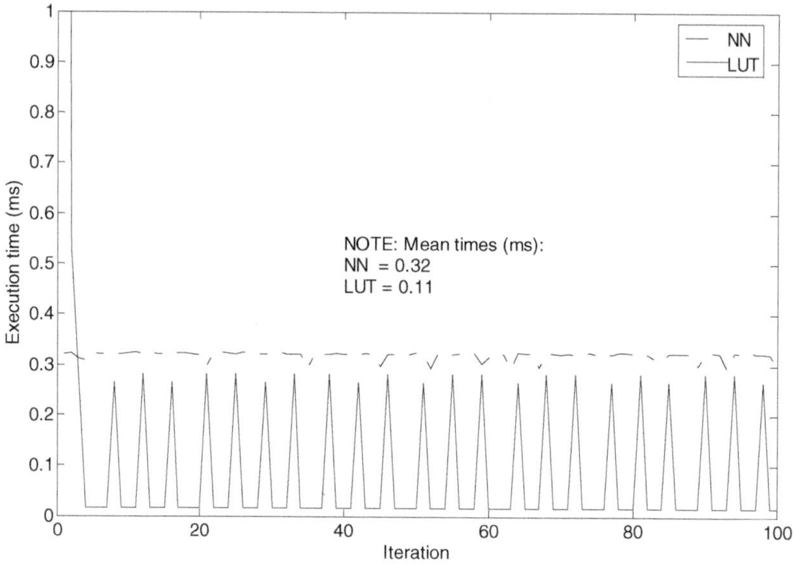

Fig. 7.36 Mean execution times: Mean taken for all 'test' data and then repeated for 100 iterations

Table 7.5 Percentage increase in estimation error when one of the ports has failed and is replaced with neighbouring port

Faulty pressure port	NN (%)			LUT (%)		
	Alpha	**Speed**	**Static Pressure**	**Alpha**	**Speed**	**Static Pressure**
P1	14.82	110.05	0.88	0.00	154.97	0.00
P2	15.50	15.62	-0.22	68.45	0.00	0.00
P3	465.27	241.14	18.38	0.00	183.75	24.45
P4	588.24	379.27	4.81	336.69	0.00	0.00
P5	46.25	145.63	-0.40	0.00	0.00	5.64
Mean (%)	**226.0 %**	**178.4 %**	**4.7 %**	**81.0 %**	**67.7 %**	**6.0 %**

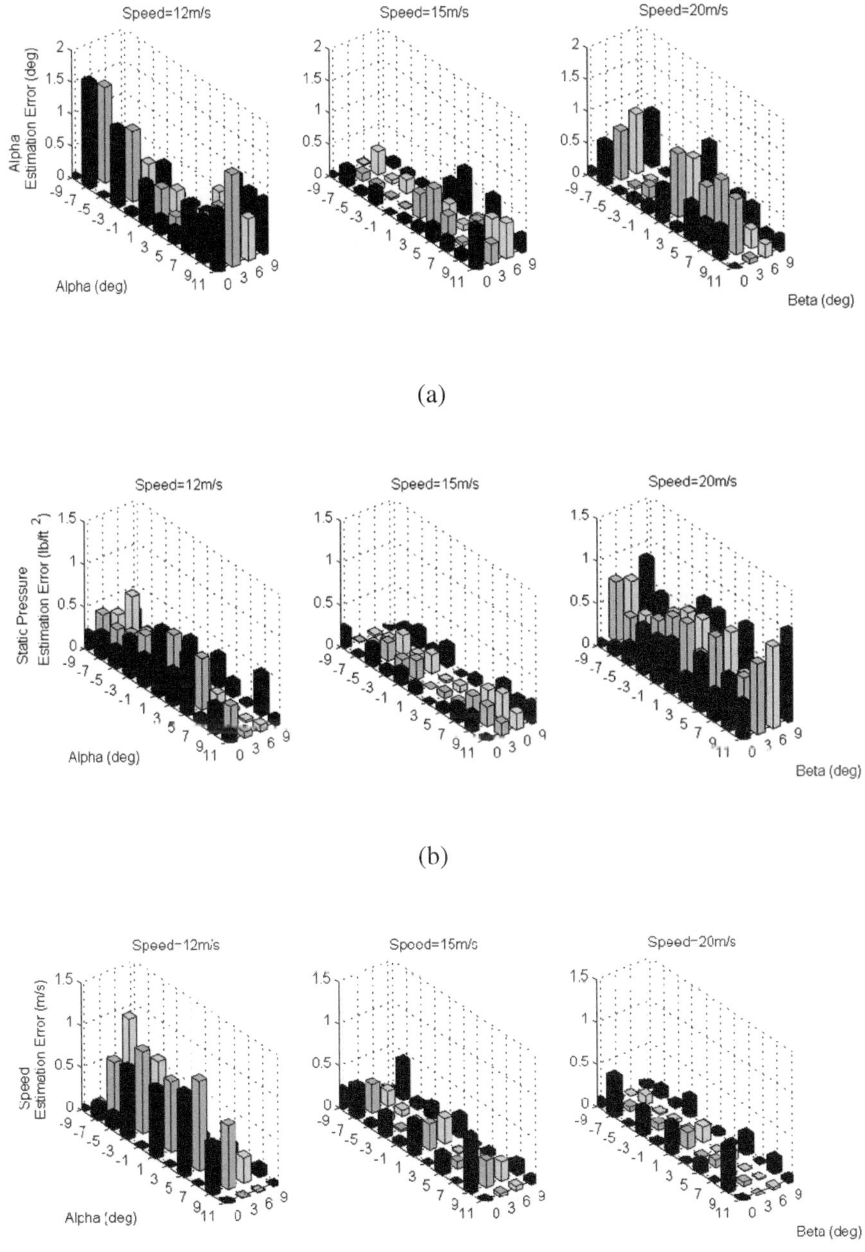

Fig. 7.37 NN estimation errors at different wind tunnel setting: (a) Alpha estimation errors (b) Static pressure estimation errors (c) Speed estimation errors. (Note only positive β range shown here)

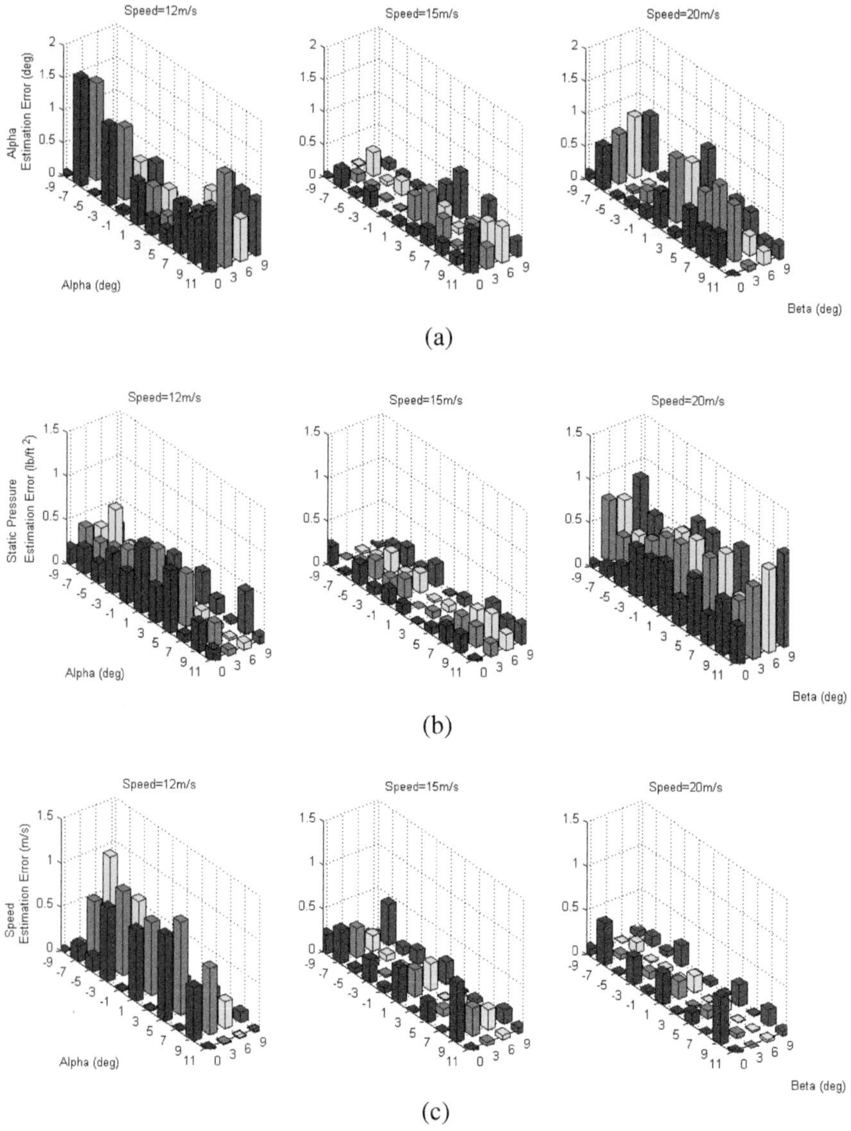

Fig. 7.38 LUT estimation errors at different wind tunnel setting: (a) Alpha estimation errors (b) Static pressure estimation errors (c) Speed estimation errors. (Note only positive β range shown here)

Conclusions

- A FADS system was designed for a MAV and tested in a wind tunnel to estimate the angle of attack, static pressure and airspeed $(P_\infty, V_\infty, \alpha)$.

- The FADS system included 5 pressure orifices placed at the wing leading edge.
- The CFD results were shown to match the wind tunnel tests. However some differences were present due to assumptions considered in the CFD simulations. This was not a problem as the CFD simulations were only used to approximate the pressure gradients on the MAV wing.
- A NN approach is used to model the aerodynamic relationship between vehicle surface pressure and air data to avoid the slow executions times associated with conventional lookup table approaches and the complexity associated with nonlinear regression methods implemented in [33, 40, 41].
- NN training was terminated according to two stopping criteria. Criterion 1 considered the *rate of change of the RMS training estimation error* (ΔRMS). If the ΔRMS remained below 0.1% for more than 100 consecutive epochs, the NN is said to have converged and training is stopped. Criterion 2 considered the increase in the RMS testing estimation error. If this error increased for more than 100 consecutive epochs, training is stopped to avoid over-fitting the NN.
- The stopping criteria were satisfied after 582 epochs of NN training resulting in a 5-3-3 EMRAN RBF NN.
- Air data RMS estimation accuracies of 0.44 lb/ft^2, 0.62 m/s and 0.51° for $\hat{P}_\infty, \hat{V}_\infty, \hat{\alpha}$ respectively were achieved for static tests where $\dot{\alpha} = \dot{\beta} = \dot{V} = 0$.
- A fault accommodation test on the 5-3-3 RBF NN revealed that a 50% reduction in NN estimation errors is possible if redundant pressure ports were considered. Similar tests showed that a 70% reduction is possible if an autoassociative NN was used instead.
- Dynamic tests (where $\dot{\alpha} \neq 0$) were carried out in a wind tunnel and a 5-3-3 NN structure was used to estimate α. An RMS estimation accuracy of 0.58° was achieved.
- A simple logic test showed that the NN estimation error is large when NN input patterns lie outside the NN domain of validity (defined here by the maximum and minimum limits of the training data set). This observation proved the importance of carefully selecting the NN training data set to avoid NN extrapolation.
- The FADS system model was also implemented as three separate LUTs. In general it was found that the LUT has faster execution times and was more robust to faults in comparison to the 5-3-3 NN. However this was only the case for a simple test where only one sensor can be faulty. On the other hand it was pointed out that LUTs are at most 3D (i.e. 2 inputs), while it is not uncommon to find NNs with more than 50 inputs. This can improve the fault tolerance capabilities of the NN especially if there are multiple sensor faults. Furthermore while a NN is implemented in a few lines of code, the LUT requires high memory usage which may not be available in our MAV.
- The FADS system designed here was insensitive to changes in sideslip. Solutions to this problem will be discussed in Chapter 8.

UAVs are currently ineligible for a standard airworthiness certificate, and are only assigned a special airworthiness certificate in the experimental category for research and development purposes [159]. However it is highly feasible that these restrictions will eventually be removed and UAVs will be integrated into the National Airspace System (NAS). One of the policies that for example the FAA adopts for regulators to issue an airworthiness certificate is based on the air-vehicle's *potential to do damage*. This categorises the air-vehicles in terms of weight, size, speed etc. Ultimately weight has relevance for airworthiness risks, and the FADS system suggested here takes this into consideration. Our FADS system weighed approximately 35g while the mini air data boom typically used by BBSR Ltd for their UAVs, weighs 170g (which can be too heavy for the MAV used in this book, as suggested by the engineers at BBSR Ltd). In this case, a reduction in weight of 135g may not seem significant, but relatively speaking, an 80% reduction in weight can be crucial in large unmanned air vehicles for both flight and airworthiness purposes. In addition, the FADS system's overall cost is almost £75 (£15 for each pressure sensor) in comparison to the air data boom which costs almost £2500. This large cost reduction is mainly of benefit to the military industry where UAVs are more likely to be destroyed during mission.

Chapter 8
Conclusions and Future Work

This book aims to exploit existing aircraft technologies to reduce costs in UAVs. The technologies include a NN-based SFDIA scheme tested on a nonlinear UAV model, and a FADS system tested on a MAV.

In industry, sensor faults are generally detected based on physical redundancy and/or limit value checking techniques. However such methods can suffer from high instrumentation costs, slow fault detection times and a high sensitivity to sensor noise. Over the years model-based SFDIA schemes have been proposed to overcome the drawbacks of traditional SFDIA methods. However the theory has generally targeted linear, fixed model based methods. Unfortunately such methods can be limited to linear, time-invariant (LTI) systems. Novel methods, as suggested by the survey carried out in [8], consider the use of NNs due to their nonlinear and adaptive structures. Fault detection techniques have been applied to large manned aircrafts [24-26, 143, 160], underwater vehicles [161], and autonomous helicopters [142], while few have been extended to fixed wing UAVs. Work carried out using NN-based methods includes [19-23, 24-28]. The work presented in this book is distinct from previous research in that a NN-based SFDIA scheme is tested on a UAV application. Model-based methods are an invaluable alternative to traditional approaches (such as physical redundancy) especially for UAVs due to weight and cost restrictions. The conclusions drawn from this book are as follows:

- The online training capabilities of NNs make them superior to most fixed, model-based approaches (such as EKFs) in terms of robustness to system and measurement noise. However too high a learning rate can cause the NN to learn the faults and therefore increase the number of undetected faults. In our case a learning rate of 0.0007 for an EMRAN RBF NN was found suitable.
- The EMRAN RBF NN was chosen due to its good generalisation capabilities and more importantly due to its ability to adapt its structure so that a minimum number of hidden neurons are used.
- The performance of a NN-based SFDIA scheme greatly depends on the residual structure implemented. In general a trade-off is required (when tuning the residual) in terms of the need to reduce the false alarms and the need to increase sensitivity to incipient faults.
- A novel residual generator referred to as RGPE was proposed to dampen the residual noise (caused by system and measurement noise). The method has proved successful as it reduced the false alarms in comparison to a

I. Samy and D.-W. Gu: Fault Detection and Flight Data Measurement, LNCIS 419, pp. 159–163.
springerlink.com

traditional residual generator (RGE). Furthermore by damping the residual noise we were able to amplify the residual and subsequently reduce the number of undetected faults.

- It was noted that residual padding may not perform as expected in the event of intermittent failures. Moreover if the minimum value in the data window to be padded, is not close-to-zero, then residual padding may not sufficiently reduce the residual average as desired.
- The RGPE approach must be carefully tuned in order to avoid damping the fault effects on the residual. In our case a residual averaging size of 50 samples, and a padding size of 50 samples, were found most suitable.
- The proposed NN-based SFDA managed to achieve the following when tested on a nonlinear UAV model; zero false alarm rate, zero undetected faults, 1.33s fault detection time, fault accommodation error of 0.41 deg/s (pitch gyro) and 0.69ms processing time per data sample (flight data sampling time was set at 20ms).
- To consider a more realistic application, multiple sensor fault scenarios were tested. The NN-based SFDIA scheme has a similar structure to the well-known GOS scheme with one NN model dedicated to only one sensor. The proposed NN-based SFDIA scheme was designed to detect faults in the pitch gyro, angle of attack sensor and the normal accelerometer.
- The NN-based SFDIA managed to achieve the following when tested on a nonlinear UAV model; 1.53s fault detection times, fault accommodation errors of 1.65 deg/s, 0.71deg, 0.88 m/s^2 for q-NN, α-NN and a_z -NN respectively, 11 false alarms, 2 undetected faults and a maximum processing time per data sample of 0.55 ms (flight data sampling time was set at 20ms).
- The NN-based SFDIA scheme was found to be highly sensitive to the fault detection time, due to its interconnected structure. Large fault detection times can result in permanent NN contamination while incipient faults result in only temporary NN contamination.
- As expected, step-type and constant bias faults are detected much quicker than incipient faults. However one observation which could not be made from the single fault tests was that step-type faults can severely damage the NN structures in the SFDIA scheme. This in turn can increase the false alarm rates. To avoid this, the fault detection time must be much lower than the results obtained here. A solution to this problem is to redesign the NNs so that they are less sensitive to faults seen in its input set e.g. by increasing the memory storage for each NN input parameter and/or increasing the number of NN input parameters.
- The study carried out here has confirmed the feasibility of using a NN-based SFDIA scheme on a UAV application. The NN processing time was on average 97% lower than the flight data sampling time (when implemented on 1.6 GHz Pentium processor). The NN online training capabilities allowed them to suitably adapt to the non-stationary flight dynamics. However it was noted that abrupt faults (such as step-type and constant bias faults) can severely damage the NN-based SFDIA performance.

The second part of this book investigates the application of a FADS system to a MAV. MAVs can be found within the spectrum of UAVs and are categorised by their low costs and weight. Traditionally air data such as airspeed and angle of attack are measured using air data booms protruding from the aircraft local flow fields. However air data booms can be too expensive and heavy for use in MAVs. As an alternative we investigate the use of a FADS system. FADS systems make use of cheap off-the-shelf pressure sensors, to convert aircraft surface pressure to air data and are an invaluable alternative to air data booms, especially for MAV applications. The concept of a FADS system is not new and has been implemented by several research groups [29-45]. However, as far as the author is aware, the FADS system has not yet been tested on MAVs (only 488mm wing span and flies at speeds as low as 8m/s). The conclusions drawn from this book are as follows:

- Traditionally the aerodynamic model used to relate the aircraft surface pressure to the air data is derived based on several assumptions (such as spherical nose shapes). Furthermore the model is highly nonlinear and can be difficult to solve. Instead, a NN model was proposed. A 5-3-3 EMRAN RBF NN was designed and shown to give estimation accuracies of 0.44 lb/ft^2, 0.62 m/s and 0.51° for $\hat{P}_\infty, \hat{V}_\infty, \hat{\alpha}$ respectively.
- The ideal pressure port locations were first investigated via 2D CFD simulations and it was found that the wing leading edge was suitable for mounting the FADS system.
- The FADS system has reduced instrumentation costs and weight by almost 97% and 80% respectively, in comparison to the air data boom used for our MAV.
- The robustness of the NN-based FADS system was investigated for faults in the pressure ports (e.g. due to port blockage, electrical wiring failure). It was found that an autoassociative-NN can significantly improve the fault accommodation performance of the FADS system in comparison to traditional methods which make use of redundant pressure ports.
- In this book, the NN training data was chosen for ease of presentation. However it was found that the NN must be robustly trained and ideally the outermost values of flight data range should be included in the NN training set. This is best implemented using real flight data as wind tunnel tests are limited to specific flight conditions.
- On average the NN processing time per data sample (0.32ms) was much lower than the flight data sampling time (20ms).
- The FADS system designed here is insensitive to changes in the sideslip. This was confirmed in the wind tunnel tests and the CFD simulations. Solutions to this will be discussed below.

The work carried out in this book has confirmed the feasibility of using NN-based SFDIA schemes and FADS systems in UAV applications. However future work must be carried out in order to validate and/or extend the work carried out so far. The future research directions are as follows:

- A limitation of the work carried out in Chapter 5, is that parameter uncertainties are only considered in the EKF equations and not the UAV model. In real applications, parameter uncertainties are likely to be present in the UAV model, and therefore it is important that the NN is tested for its robustness to such uncertainties.
- The flight conditions considered in the SFDA scheme, mainly considered 3-2-1-1 elevator input demands, as a first step towards analysing the NN performance. However future work must consider more realistic flight scenarios.
- Prior to fault detection, the NN-SFDIA scheme is simply a health monitoring system, i.e. a fault alarm system. However once the fault is detected, the NN estimates must replace the faulty sensor. In this case, the NN estimates can be used in the control feedback loops. Therefore the stability of the control system to NN estimations must be investigated.
- Real flight data can help validate the robustness of the NN and RGPE structures to system and measurement noise. Artificial faults can be added to the sensor data and the sensitivity of the NN-based SFDIA scheme to incipient faults can be further investigated. Furthermore, multiplicative faults (we have mainly considered additive faults) can be considered.
- The RGPE method suggested here assumes that faults have a permanent effect on the residual. However intermittent failures are not considered. It is important to investigate the sensitivity of RGPE to intermittent failures if it is to be applied in a real system.
- The FADS system can be implemented on both wings of the MAV. We can then average the air data estimations from both sets. This way we can improve the fault tolerance capabilities and sensitivity to noise of the overall FADS system.
- The MAV (instead of just the wing) can be tested in a wind tunnel prior to any flight tests.
- The FADS system must be tested in real flight to gain more confidence in their estimations accuracies, execution speeds and stability to fluctuating pressure measurements (caused by e.g. atmospheric debris partially blocking the pressure ports). The FADS system can be flight tested in parallel to an air data boom and the latter can be used to validate the performance of the FADS system.
- In Chapter 7 we suggested that wind tunnel data is not ideal to train the NN, for several reasons; 1) Altitude cannot be changed 2) Certain flight conditions (e.g. high angle of attack rates) could not be investigated 3) Environmental conditions experienced during real flight (e.g. poor weather conditions) cannot be considered in the wind tunnel. These three conditions are extremely important if the NN is to be robustly trained and therefore real flight data is required.

- If the FADS system is successful in real flight, then we can investigate the possibility of combining the FADS system and air data boom for a more accurate, fault tolerant and robust (to sensor noise) air data system. Furthermore the air data boom can be used to train (online) the NN-FADS system. A suitable air data boom (worth approximately £600) has already been purchased from SpaceAge Control and has been mounted on the MAV. The air data measurements from the boom, can therefore be used to verify the performance of the FADS system.
- The study presented in this book has investigated the fault accommodation performance of the NN-based FADS system. However we have not yet investigated the fault detection performance. This is important, as fault accommodation is only possible if the fault is detected. We can use our knowledge of SFDIA (Chapters 5-6) to design a SFDIA scheme for the FADS system.
- As pointed out in Chapter 7, estimating the sideslip (with our current design) can be difficult. However, from the 3D CFD simulations, it was found that pressure close to the wing root and wing tip is sensitive to sideslip. Instead of re-designing the entire FADS system, we can simply add extra pressure ports so that sideslip can be estimated. This is feasible for several reasons; 1) There is sufficient space and weight left on the MAV to mount extra pressure sensors, 2) The NN processing time is currently 98% lower than the flight data sampling time and therefore increasing the number of pressure ports should not cause any significant time delays 3) The 3D CFD simulations in Chapter 7 show that the pressure varies almost linearly with sideslip for ports located close to the wing leading edge 4) Pressure sensors are cheap (£15 each) and therefore increasing the number of pressure ports will not be costly. Another approach which could be implemented to estimate sideslip, is if we place a pressure port at the tip of each wing and take the differential pressure of the two pressure measurements.

References

[1] Wong, K.C.: Aerospace industry opportunities in Australia-unmanned aerial vehicles (UAVs). Department of Aeronautical Engineering, University of Sydney (2007)

[2] Wong, K.C., Bil, C., Gordon, D., Gibbens, P.W.: Study of the unmanned aerial vehicle (UAV) market in Australia. Department of Aeronautical Engineering, University of Sydney (1997)

[3] Buonanno, A., Cook, M.V.: An aerodynamic simulation model of the Eclipse UAV, Internal Report FLAV-A01-002 (April 29, 2005)

[4] Buonanno, A., Cook, M.V.: Flight dynamics model of the flying demonstrator UAV, Internal Report FLAV-A01-003 (June 13, 2005)

[5] Whidborne, J.F., Cowling, I.D., Yakimenko, O.A.: A direct method for UAV guidance and Control. In: 23rd International Conference on Unmanned Air Vehicle Systems, Bristol, UK, pp. 37.1–37.13 (April 2008)

[6] Cooke, A.K., Cowling, I.D., Erbsloeh, S.D., Whidborne, J.F.: Low cost system design and development towards an autonomous rotor vehicle. In: 22nd International Conference on Unmanned Air Vehicle Systems, Bristol, UK, 28.1–28.9 (2007)

[7] La Franchi, P.: Grand designs: An EC-funded research project has unveiled its proposals for a new generation of aircraft that are intended to give Europe the edge in the civil UAV sector. In: Flight International, pp. 109–114 (2005)

[8] Isermann, R., Balle, P.: Trends in the applications of model-based fault detection and diagnosis of technical processes. Control Engineering Practice 5(5), 709–719 (1997)

[9] Patton, R.J., Frank, P.M., Clark, R.N.: Fault diagnosis in dynamic systems-theory and applications. Prentice Hall, London (1989)

[10] Gertler, J.: Survey of model-based failure detection and isolation in complex plants. IEEE Control Systems Magazine 8, 3–11 (1988)

[11] Chen, J., Patton, R.J.: Robust model-based fault diagnosis for dynamic systems. Kluwer Academic Publishers, USA (1999)

[12] Frank, P.M.: Fault diagnosis in dynamic systems using analytical and knowledge-based redundancy-A survey and some new results. Automatica 26(3), 459–474 (1990)

[13] Willsky, A.S.: A survey of design methods for failure detection in dynamic systems. Automatica 12(6), 601–611 (1976)

[14] Isermann, R.: Model based fault detection and diagnosis- Status and applications. In: 16th Symposium on Automatic Control in Aerospace, St. Petersburg (2004)

[15] Betta, G., Pietrosanto, A.: Instrument fault detection and isolation: State of the art and new research trends. IEEE Transactions on Instrumentation And Measurement 49(1), 100–107 (2000)

[16] Isermann, R.: Supervision, fault detection and fault diagnosis methods. An introduction. Control Engineering Practice 5(5), 639–652 (1997)

[17] Simani, S., Fantuzzi, C., Patton, R.J.: Model-based fault diagnosis in dynamic systems using identification techniques. Springer, London (2003)

[18] Gertler, J.: Fault detection and diagnosis in engineering systems. Marcel Dekker Inc., New York (1998)

[19] Perla, R., Mukhopadhyay, S., Samantam, A.N.: Sensor fault detection and isolation using artificial neural networks. In: IEEE Region Conference TENCON, vol. 4, pp. 676–679 (2004)

[20] Fernandes, R.G., Silva, D., Oliveira, L., Neto, A.: Faults detection and isolation based on neural networks applied to a levels control system. In: International Joint Conference on Neural Networks, Florida, USA (2007)

[21] Capriglione, D., Liguori, C., Pietrosanto, A.: Real-time implementation of IFDIA scheme in automotive systems. IEEE Transactions on Instrumentation and Measurement 56(3), 824–830 (2007)

[22] Li, R., Olson, J.H., Chester, D.L.: Dynamic fault detection and diagnosis using neural networks. In: Proceedings of the IEEE International Symposium in Intelligent Control, vol. 2, pp. 1169–1174 (1990)

[23] Naidu, S., Zafiriou, E., McAvoy, T.J.: Use of neural networks for sensor failure detection in a control system. IEEE Control Systems Magazine 10(3), 49–55 (1990)

[24] Campa, G., Fravolini, M.L., Napolitano, M., Seanor, B.: Neural networks based sensor validation for the flight control system of a B777 research model. In: Proceedings of the American Control Conference, vol. 1, pp. 412–417 (2002)

[25] Napolitano, M., An, Y., Seanor., B., Pispistos, S., Martinelli, D.: Application of a neural sensor validation scheme to actual Boeing B737 flight data. In: Proceedings of the AIAA Guidance Navigation and Control Conference (1999)

[26] Napolitano, M., An, Y., Seanor, B.: A fault tolerant flight control system for sensor and actuator failures using neural networks. Aircraft Design 3, 103–128 (2000)

[27] Napolitano, M., Neppach, C., Casdorph, V., Naylor, S., Innocenti, S., Silvestri, M.: Neural network based scheme for sensor failure detection identification and accommodation. Journal of Guidance, Control and Dynamics 18(6), 1280–1286 (1995)

[28] Fravolini, M., Campa, G., Napolitano, M., Perhinschi, M.: Learning based sensor validation scheme within flight control laws. Journal of Guidance, Control and Dynamics 27(2), 307–310 (2004)

[29] Cary, J.P., Keener, E.R.: Flight evaluation of the X-15 Ball-Nose Flow -Direction sensor as an airdata system, NASA TN D-2923 (1965)

[30] Wolowicz, C.H., Gosett, T.D.: Operational and performance characteristics of the X-15 spherical hypersonic flow direction sensor, NASA TN D-3076 (1965)

[31] Larson, T.J., Siemers III, P.M.: Subsonic tests of an All-Flush-Pressure-Orifice air data system, NASA TP-1871 (1981)

[32] Larson, T.J., Whitmore, S.A., Ehernberger, L.J., Johnson, J.B., Siemers III, P.M.: Qualitative evaluation of a flush air data system at transonic speeds and high angles of attack, NASA TP-2716 (1987)

[33] Larson, T.J., Moes, T.R., Siemers III, P.M.: Wind tunnel investigation of a flush airdata system at Mach numbers from 0.7 to 1.4, NASA TM-101697 (1990)

[34] Whitmore, S.A., Davis, R.J., Fife, J.M.: In flight demonstration of a real time flush airdata sensing system, NASA TM-104314 (1995)

[35] Rohloff, T.: Development and evaluation of neural network flush air data sensing systems, PhD book, Department of Mechanical Engineering, University of California (1998)

[36] Crowther, W.J., Lamont, P.J.: A neural network approach to the calibration of a flush air data system. Aeronautical Journal 105(1044), 85–95 (2001)

[37] Brown, E.N., Friehe, C.A., Lenschow, D.H.: The use of pressure fluctuations on the nose of an aircraft for measuring air motion. Journal of Climate and Applied Meteorology 22, 171–180 (1983)

[38] Whitmore, S.A., Moes, T.R., Czerniejewski, M.W., Nichols, D.A.: Application of a Flush Airdata Sensing System to a Wing leading edge (LE-FADS), NASA TM-104267 (1993)
[39] Siemers III, P.M., Wolf, H., Henry, M.W.: Shuttle Entry Air Data System (SEADS)-Flight Verification of an Advanced Airdata System Concept, AIAA-88-2104 (1998)
[40] Whitmore, S.A., Stephen, A., Moes, T.R., Timothy, R., Larson, T.J.: Preliminary Results From a Subsonic High Angle-of-Attack Flush Airdata Sensing (HI-FADS) System: Design, Calibration, and Flight Test Evaluation, NASA TM-101713 (1990)
[41] Whitmore, S.A., Moes, T.R.: Failure Detection and Fault Management Techniques for a Pneumatic High-Angle-of-Attack Flush Airdata Sensing (HI-FADS) System, NASA TM-4335 (1992)
[42] Cobleigh, B.R., Whitmore, S.A., Haering, E.A., Borrer, J., Roback, V.E.: Flush Airdata Sensing (FADS) system calibration procedures and results for blunt forebodies, NASA TP-209012 (1999)
[43] Whitmore, S.A., Cobleigh, B.R., Haering, E.A.: Design and Calibration of the X-33 Flush Airdata Sensing (FADS) System, NASA TM-206540 (1998)
[44] Rediniotis, O., Vijayagopal, R.: Miniature multihole pressure probes and their neural network based calibration. AIAA Journal 37(6), 666–674 (1999)
[45] Rediniotis, O., Chrysanthakopoulos, G.: Application of neural networks and fuzzy logic to the calibration of the seven-hole probe. Journal of Fluids Engineering-Transactions of the ASME 120(1), 95–101 (1998)
[46] Samy, I., Postlethwaite, I., Gu, D.: Subsonic tests of a flush air data sensing system applied to a fixed-wing micro air vehicle. In: Unmanned Aircraft Systems, pp. 275–295. Springer, Netherlands (2009)
[47] Samy, I., Postlethwaite, I., Gu, D.: Neural network sensor validation scheme demonstrated on a UAV model. In: IEEE Proceedings of CDC, Cancun, Mexico, pp. 1237–1242 (December 2008)
[48] Samy, I., Postlethwaite, I., Gu, D.: SFDIA of consecutive sensor faults using neural networks- demonstrated on a UAV. International Journal of Control 83(11), 2308–2327 (2010)
[49] Samy, I., Postlethwaite, I., Gu, D.: Sensor fault detection and accommodation using neural networks with application to a non-linear unmanned air vehicle model. Proceedings of IMechE Part G: Journal of Aerospace Engineering 224(4), 437–447 (2010)
[50] Samy, I., Postlethwaite, I., Gu, D.: Survey and application of sensor fault detection and isolation. Control Engineering Practice (Article in press, 2011)
[51] Samy, I., Postlethwaite, I., Gu, D., Green, J.: EMRAN RBF NN based flush air data sensing system-demonstrated on a mini air vehicle. AIAA Journal of Aircraft 47(1), 18–31 (2010)
[52] Favre, C.: Fly-by-wire for commercial aircraft: the Airbus experience. International Journal of Control 59(1), 139–157 (1994)
[53] Gilmore, J., Mckern, R.: A redundant strap-down inertial system mechanization-SIRU. In: AIAA Guidance, Control and Flight Mechanics Conference, California, USA (1970)
[54] Cikanek III, H.A.: Space shuttle main engine failure detection. IEEE Control Systems Magazine 6(3), 13–18 (1986)
[55] Carden, E.P.: Vibration based condition monitoring: A review. Structural Health Monitoring 3(4), 355–377 (2004)
[56] Hakami, B., Newborn, J.: Expert systems in heavy industry: An application of ICLX in a British a steel corporation works. ICL Technical Journal 3(4), 347–359 (1983)
[57] Kumamoto, H., Ikenchi, K., Inoue, K., Henley, E.J.: Application of expert system techniques to fault diagnosis. The Chemical Engineering Journal 29(1), 1–9 (1984)

[58] Isermann, R.: Process fault detection based on modeling and estimation methods: A survey. Automatica 20, 387–404 (1984)
[59] Venkatasubramanian, V., Rengaswamy, R., Yin, K., Kavuri, S.N.: A review of process fault detection and diagnosis Part I: Quantitative model based methods. Computers and Chemical Engineering 27, 293–311 (2003)
[60] Poon, F.W.: Observer based robust fault detection: Theory and Rolling mill Case study, PhD book, Department of Engineering. University of Leicester (2000)
[61] Patton, R.J.: Fault-tolerant control. The 1997 situation. In: IFAC SAFEPROCESS 1997, vol. 2, pp. 1033–1055 (1997)
[62] Blanke, M., Patton, R.J.: Industrial actuator benchmark for fault detection and isolation. Control Engineering Practice 3(12), 1727–1730 (1995)
[63] Chow, E.Y., Willsky, A.S.: Analytical redundancy and the design of robust detection systems. IEEE Transactions Automatic Control 29(7), 603–614 (1984)
[64] Tan, C.P.: Sliding mode observers for fault detection and isolation, PhD book, Department of Engineering, University of Leicester (2002)
[65] Patton, R.J., Willcox, S.W., Winter, J.S.: A parameter insensitive technique for aircraft sensor fault analysis. Journal of Guidance Control and Dynamics 10(3), 359–367 (1987)
[66] Edwards, C., Spurgeon, S.K.: On the development of discontinuous observers. Int. Journal of Control 59(5), 1211–1229 (1994)
[67] Lou, X., Willsky, A.S., Verghese, G.: Optimal robust redundancy relations for failure detection in uncertainty systems. Automatica 22(3), 333–344 (1986)
[68] Mironovski, L.A.: Functional diagnosis of linear dynamic systems. Automn Remote Control 41, 1122–1143 (1979)
[69] Desai, M., Ray, A.: A fault detection and isolation methodology-theory and application. In: American Control Conference, San Diego, California, USA (1984)
[70] Massoumnia, M.A., Velde, W.: Generating parity relations for detecting and identifying control system component failures. Journal of Guidance, Control and Dynamics 11(1), 60–65 (1988)
[71] Patton, R.J., Chen, J.: A review of party space approaches to fault diagnosis. In: Preprints of IFAC/IMACS Symposium: SAFEPROCESS 1991, Baden-Baden, vol. 1, pp. 239–255 (1991)
[72] Gertler, J.: Analytical redundancy methods in failure detection and isolation. In: Preprints of IFAC/IMACS Symposium: SAFEPROCESS 1991, Baden-Baden, vol. 1, pp. 9–21 (1991)
[73] Staroswiecki, M., Cassar, J.P., Cocquempot, V.: Generation of optimal structured residuals in the parity space. In: Preprints of the 12th IFAC World Congress, Australia, Vol. 8, pp. 299–305 (1993)
[74] Ding, X., Guo, L., Jeinsch, T.: A characterization of parity space and its application to robust fault detection. IEEE Transactions on Automatic Control 44(2), 337–343 (1999)
[75] Luenberger, D.J.: An introduction to observers. IEEE Transactions on Automatic Control 16(6), 596–602 (1971)
[76] Kalman, R.E.: A new approach to linear filtering and prediction problems. Journal of Basic Engineering 82, 35–45 (1960)
[77] Clark, R.N.: A simplified instrument failure detection scheme. IEEE Transactions on Aerospace and Electronic Systems 14(4), 558–563 (1978)
[78] Clark, R.N., Setzer, W.: Sensor fault detection in a system with random disturbances. IEEE Transactions on Aerospace and Electronic Systems 16(4), 468–473 (1980)
[79] Clark, R.N.: Instrument fault detection. IEEE Transactions on Aerospace and Electronic Systems 14(3), 456–465 (1978)

[80] Frank, P.M.: Advanced fault detection and isolation schemes using nonlinear and robust observers. In: 10th IFAC Congress on Automatic Control, Munchen, vol. 3, pp. 63–68 (1987)

[81] Mehra, R.K., Peschon, J.: An innovations approach to fault detection and diagnosis in dynamic systems. Automatica 7, 637–643 (1971)

[82] Willsky, A.S., Jones, H.L.: A generalized likelihood approach to state estimation in linear systems subjected to abrupt changes. In: Proceedings of CDC, Arizona, USA (1974)

[83] Willsky, A.S., Jones, H.L.: A generalized likelihood ratio approach to the detection and estimation of jumps in linear systems. IEEE Transactions on Automatic Control 21, 108–121 (1976)

[84] Willsky, A.S., Deyst, J.J., Crawford, B.S.: Adaptive filtering and self-test methods for failure detection and compensation. In: American Control Conference, Austin, USA (1974)

[85] Montgomery, R.C., Caglayan, A.K.: Failure accommodation in digital flight control systems by Bayesian decision theory. Journal of Aircraft 13(2), 69–75 (1976)

[86] Tzafestas, S.G., Watanabe, K.: Modern approaches to system/sensor fault detection and diagnosis. Journal A 31(4), 42–57 (1990)

[87] Basseville, M.: Information criteria for residual generation and fault detection and isolation. Automatica 33(5), 783–803 (1997)

[88] Eide, P., Maybeck, B.: An MMAE failure detection system for the F-16. IEEE Transactions on Aerospace and Electronic Systems 32(3), 1125–1136 (1996)

[89] Berec, L.: A multi-model method to fault detection and diagnosis: Bayesian solution. An introductory treatise. International Journal of Adaptive Control and Signal Processing 12(1), 81–92 (1998)

[90] Shapiro, E.Y., Decarli, H.E.: Analytical redundancy for flight control on the Lockheed L-1011 Aircraft. In: Proceedings of CDC, San Diego, USA (1979)

[91] Deyst, J.J., Deckert, J.C.: Maximum likelihood failure detection techniques applied to the shuttle RCS Jets. Journal of Spacecraft and Rockets 13, 65–74 (1976)

[92] Beard, R.V.: Failure accommodation in linear systems through self reorganization, PhD book, Massachusetts Institute of Technology, USA (1971)

[93] Jones, H.L.: Failure detection in linear systems, PhD Book, Department of Aeronautics. Massachusetts Institute of Technology, USA (1973)

[94] Young, P.C.: Parameter estimation for continuous time models-a survey. Automatica 17(1), 23–29 (1981)

[95] McCulloch, W.S., Pitts, W.: A logical calculus of the ideas immanent in nervous activity. Bulletin of Mathematical Biophysics 5, 115–133 (1943)

[96] Minsky, M.L., Papert, S.A.: Perceptrons. MIT Press, Cambridge (1969)

[97] Haykin, S.: Neural networks: a comprehensive foundation. Macmillan College Publishing Company, USA (1994)

[98] Minsky, M.L.: Steps towards artificial Intelligence. Proceedings of the Institute of Radio Engineers 49, 8–30 (1961)

[99] Skogestad, S., Postlethwaite, I.: Multivariable feedback control analysis and design. John Wiley & Sons Ltd., West Sussex (2005)

[100] Hopkins, J.C., Himmelblau, D.M.: Artificial neural network models for knowledge representation in chemical engineering. Computers Chemical Engineering 12(9/10), 881–890 (1988)

[101] Venkatasubramanian, V., Chan, K.: A neural network methodology for process fault diagnosis. AIChE Journal 35(12), 1993–2001 (1989)

[102] Hoskins, J.C., Kaliyur, K.M., Himmelblau, D.M.: Fault diagnosis in complex chemical plants using artificial neural networks. AIChE Journal 37(1), 137–141 (1991)

[103] Beale, R., Jackson, T.: Neural computing: an introduction. IOP Publishing Ltd., Bristol (1992)

[104] Fausett, L.: Fundamentals of neural networks: architectures, algorithms and applications. Prentice Hall International Inc., New Jersey (1994)

[105] Watanabe, U., Himmelblau, D.M.: Instrument fault detection in systems with uncertainties. International Journal of System Science 13, 137–158 (1982)

[106] Emami-Naeini, A.E., Akhter, M.M., Rock, S.M.: Effect of model uncertainty on failure detection: the threshold selector. IEEE Transactions on Automatic Control 33(2), 1106–1115 (1988)

[107] Djeziri, M.A., Aitouche, A., Bouamama, B.: Sensor fault detection of energetic system using modified parity space approach. In: Proceedings of CDC, New Orleans, LA, USA (2007)

[108] Chan, C.W., Hua, S., Yue, Z.: Application of fully decoupled parity equation in fault detection and identification of DC motors. IEEE Transactions on Industrial Electronics 53(4), 1277–1284 (2006)

[109] Schneider, S., Weinhold, N., Ding, S.X., Rehm, A.: Parity space based FDI scheme for vehicle lateral dynamics. In: IEEE Conference on Control Applications, Toronto, Canada (2005)

[110] Halder, P., Chaudhuri, S.K., Mukhopadhyay, S.: Online sensor fault detection, isolation and accommodation in tactical aerospace vehicle. In: IEEE Region Conference TENCON, vol. 4(21-24), pp. 684–686 (2004)

[111] Dan, W., Zhiliang, W., Yubin, Y., Xiaobing, N.: An FDI approach for aircraft actuator partial failure. In: IEEE Chinese Control Conference, Hunan, China (2007)

[112] Aouf, N., Boulet, B.: Fault diagnosis techniques: application to the thermoforming process. In: 4th International Conference on Control and Automation, Montreal, Canada (2003)

[113] Liberatore, S., Speyer, J.L., Hsu, A.: Fault detection filter applied to structure health monitoring. In: Proceedings of CDC, Hawaii, USA (2003)

[114] Jiang, T., Khorasani, K., Tafazoli, S.: Parameter estimation based fault detection, isolation and recovery for nonlinear satellite models. IEEE Transactions on Control Systems Technology 16(4), 799–808 (2008)

[115] Moseler, O., Isermann, R.: Application of model-based fault detection to a brushless DC motor. IEEE Transactions on Industrial Electronics 47(5), 1015–1020 (2000)

[116] Capriglione, D., Liguori, C., Pianese, C., Pietrosanto, A.: Online sensor fault detection, isolation and accommodation in automotive engines. IEEE Transactions on Instrumentation and Measurement 52(4), 1182–1189 (2003)

[117] Campa, G., Krishnamurty, M., Gautam, M., Napolitano, M.R., Perhinschi, M.: A neural network based sensor validation scheme for heavy-duty diesel engines. In: 14th Mediterranean Conference on Control and Automation, Ancona, Italy (2006)

[118] Fravolini, M.L., Campa, G., Napolitano, K., Song, Y.: Minimal resource allocating networks for aircraft SFDIA. In: IEEE International Conference on Advanced Intelligent Mechatronics, Como, Italy (2001)

[119] Andersen, D., Haley, D.: NASA tests new laser air data system on SR-71 Blackbird, NASA: `http://www.nasa-usa.de/home/hqnews/1993/93-163.txt` (accessed September 17, 1993)

[120] Edward, A., Haering Jr.: Airdata Measurement and Calibration, NASA TM-104316 (1995)

[121] Anderson, J.D.: Introduction to flight. McGraw Hill, USA (2008)

[122] `http://www.spaceagecontrol.com/Adpmain`

[123] SpaceAge Control Inc.: Calibration of SpaceAge Control 100400 Mini air data boom, SpaceAge Control Report X004A(NC) (2001)

[124] Anderson, J.D.: Fundamentals of Aerodynamics, 2nd edn. McGraw-Hill, USA (1991)
[125] Houghton, E.L., Carpenter, P.W.: Aerodynamics for engineering students, 5th edn. Butterworth-Heinemann, Oxford (2003)
[126] Churchland, P.S.: Neurophilosophy: Toward a unified science of the mind/brain. MIT Press, Cambridge (1986)
[127] Faro, A., Giordano, D., Spampinato, C.: Evaluation of the traffic parameters in a metropolitan area by fusing visual perceptions and CNN processing of webcam images. IEEE Transactions on Neural Networks 19(6), 1108–1129 (2008)
[128] Parisi, A., Parisi, F., Diaz, D.: Forecasting gold price changes: Rolling and recursive neural network models. Journal of Multinational Financial Management 18, 477–487 (2008)
[129] Wang, H.N., Cui, Y.M., Li, R., Zhang, L.Y., Han, H.: Solar flare forecasting model supported with artificial neural network techniques. Advances in Space Research 42, 1464–1468 (2008)
[130] Gallinari, P.: Industrial applications of neural networks. World Scientific Publishing Co. Pte. Ltd., Singapore (1998)
[131] Powell, M.J.D.: Radial basis function for multivariable interpolation: a review. In: Mason, J.C., Cox, M.G. (eds.) Algorithms for Approximation, pp. 143–167. Clarendon Press, Oxford (1987)
[132] Lu, Y., Sundararajan, N., Saratchandran, P.: Analysis of minimal radial basis function network algorithm for real-time identification of nonlinear dynamic systems. IEEE Proceedings on Control Theory and Applications 147(4), 476–484 (2000)
[133] Chen, S., Cowan, F.N., Grant, P.M.: Orthogonal least squares learning algorithm for radial basis function networks. IEEE Transactions on Neural Networks 2, 302–309 (1991)
[134] Platt, J.C.: A resource allocating network for function interpolation. Neural Computing 3, 213–225 (1991)
[135] Kadirkamanathan, V., Niranjan, M.: A function estimation approach to sequential learning with neural networks. Neural Computing 5, 954–975 (1993)
[136] Napolitano, M.R., Windon, D.A., Casanova, J.L., Innocenti, M., Silvestri, G.: Kalman filters and neural network schemes for sensor validation in flight control systems. IEEE Transactions on Control Systems Technology 6(5), 596–611 (1998)
[137] Sorenson, H.W.: Kalman Filtering: Theory and Application. IEEE Press, New York (1985)
[138] Maybeck, P.: Stochastic models, estimation and control, vol. 1. Academic Press, London (1979)
[139] Brown, R.G.: Introduction to random signal analysis and Kalman filtering. Wiley, USA (1983)
[140] Cook, M.V.: Flight dynamics principles. Arnold, Great Britain (1997)
[141] Welch, G., Bishop, G.: An introduction to the Kalman filter. University of North Carolina at Chapel Hill, NC 27599-3175 (2006)
[142] Heredia, G., Ollero, A., Mahtani, R., Remub, V., Mausial, M.: Detection of sensor faults in autonomous helicopters. In: Proceedings of the 2005 IEEE International Conference on Robotics and Automation, pp. 2229–2234 (2005)
[143] An, Y.: A design of fault tolerant flight control systems for sensor and actuator failures using on-line learning neural networks, PhD book, Department of Mechanical and Aerospace Engineering, West Virginia, University, USA (1998)
[144] MH64 airfoil: `http://www.mh-aerotools.de/airfoils/mh64koo.htm`

[145] Houghton, E.L., Carruthers, N.B.: Aerodynamics for engineering students, 3rd edn. Edward Arnold, London (1982)
[146] Larson, T.J., Flechner, S.G., Siemers, P.M.: Wind tunnel investigation of an all flush orifice air data system for a large subsonic aircraft, NASA TP 1642 (May 1980)
[147] Brown, E.N., Friehe, C.A., Lenschow, D.H.: The use of pressure fluctuations on the nose of an aircraft for measuring air motion. Journal of Applied Meteorology 22(1), 171–180 (1983)
[148] Sastry, C.V., Raman, K.S., Babu, L.B.: Failure management scheme for use in a flush air data system. Aircraft Design 4(4), 151–162 (2001)
[149] Whitmore, S.A.: Development of a pneumatic high angle of attack flush airdata sensing system. In: NASA Technical Memorandum 104241 (November 1991)
[150] Courrieu, P.: Three algorithms for estimating the domain of validity of feedforward neural networks. Neural Networks 7(1), 169–174 (1994)
[151] Helliwell, I.S., Torega, M.A., Cottis, R.A.: Accountability of neural networks trained with 'Real World' data. In: 4th International Conference on Artificial Neural Networks, pp. 218–222 (1995)
[152] Lancaster, P., Salkauskas, K.: Curve and surface fitting an introduction. Academic Press, London (1986)
[153] Phillips, G.M.: Interpolation and approximation by polynomials. Springer, New York (2003)
[154] Cohen, A., Rabut, C., Schumaker, L.L.: Curve and surface design. In: Proceedings of Conference on Approximation Theory, Saint-Malo, France, vol. 1 (July 1999)
[155] Laurent, P., Sablonniere, P., Schumaker, L.L.: Curve and surface design. In: Proceedings of Conference on Approximation Theory, Saint- Malo, France, vol. 2 (July 1999)
[156] Farrashkhalvat, M., Miles, J.P.: Basic structured grid generation with an introduction to unstructured grid generation: With an introduction to unstructured grid generation. Butterworth-Heinemann, UK (2003)
[157] Rogers, C.A.: Packing and Covering. Cambridge University Press, Cambridge (1964)
[158] Green, P.J., Sibson, R.: Computing Dirichlet tessellations in the plane. Computing Journal 21, 168–173 (1978)
[159] FAA: Unmanned Aircraft Systems (UAS) Certifications and Authorizations. US Department of Transportation (2007), http://www.faa.gov/aircraft/air_cert/design_approvals/uas/cert/ (accessed November 5, 2007)
[160] Motyka, P., Bonnice, W., Hall, S., Wagner, E.: The evaluation of failure detection and isolation algorithms for restructurable control, NASA Contractor Report 177983 (1985)
[161] Alessandri, A., Caccia, M., Veruggio, G.: Fault detection of actuator faults in unmanned underwater vehicle. Control Engineering Practice 7, 357–368 (1999)

Lecture Notes in Control and Information Sciences

Edited by M. Thoma, F. Allgöwer, M. Morari

Further volumes of this series can be found on our homepage:
springer.com

Vol. 419: Samy, I.; Gu, D.-W.:
Fault Detection and Flight Data Measurement
170 p. 2011 [978-3-642-24051-5]

Vol. 418: Alberer, D.; Hjalmarsson, H.; del Re, L.:
Identification for Automotive Systems
348 p. 2011 [978-1-4471-2220-3]

Vol. 417: Johansson, R.; Rantzer A.:
Distributed Decision Making and Control
412 p. 2011 [978-1-4471-2264-7]

Vol. 416: Varga, A.; Hansson, A.; Puyou, G.:
Optimization Based Clearance of Flight Control Laws
451 p. 2011 [978-3-642-22626-7]

Vol. 415: Chesi, G.:
Domain of Attraction
283 p. 2011 [978-0-85729-958-1]

Vol. 414: Imine, H.; Fridman, L.; Shraim, H.; Djemai, M.:
Sliding Mode Based Analysis and Identification of Vehicle Dynamics
127 p. 2011 [978-3-642-22223-8]

Vol. 413: Eleftheriou, E.; Reza Moheimani, S.O.:
Control Technologies for Emerging Micro and Nanoscale Systems
289 p. 2011 [978-3-642-22172-9]

Vol. 412: Fridman, L.; Moreno, J.; Iriarte, R.:
Sliding Modes after the first Decade of the 21st Century
XXX p. 2011 [978-3-642-22163-7]

Vol. 411: Kaczorek, T.;
Selected Problems of Fractional Systems Theory
344 p. 2011 [978-3-642-20501-9]

Vol. 410: Bourlès, H., Marinescu, B.:
Linear Time-Varying Systems
637 p. 2011 [978-3-642-19726-0]

Vol. 409: Xia, Y., Fu, M., Liu, G.-P.:
Analysis and Synthesis of Networked Control Systems
198 p. 2011 [978-3-642-17924-2]

Vol. 408: Richter, J.H.;
Reconfigurable Control of Nonlinear Dynamical Systems
291 p. 2011 [978-3-642-17627-2]

Vol. 407: Lévine, J., Müllhaupt, P.:
Advances in the Theory of Control, Signals and Systems with Physical Modeling
380 p. 2010 [978-3-642-16134-6]

Vol. 406: Bemporad, A., Heemels, M., Johansson, M.:
Networked Control Systems
appro. 371 p. 2010 [978-0-85729-032-8]

Vol. 405: Stefanovic, M., Safonov, M.G.:
Safe Adaptive Control
appro. 153 p. 2010 [978-1-84996-452-4]

Vol. 404: Giri, F.; Bai, E.-W. (Eds.):
Block-oriented Nonlinear System Identification
425 p. 2010 [978-1-84996-512-5]

Vol. 403: Tóth, R.;
Modeling and Identification of Linear Parameter-Varying Systems
319 p. 2010 [978-3-642-13811-9]

Vol. 402: del Re, L.; Allgöwer, F.; Glielmo, L.; Guardiola, C.; Kolmanovsky, I. (Eds.):
Automotive Model Predictive Control
284 p. 2010 [978-1-84996-070-0]

Vol. 401: Chesi, G.; Hashimoto, K. (Eds.):
Visual Servoing via Advanced Numerical Methods
393 p. 2010 [978-1-84996-088-5]

Vol. 400: Tomás-Rodríguez, M.; Banks, S.P.:
Linear, Time-varying Approximations to Nonlinear Dynamical Systems
298 p. 2010 [978-1-84996-100-4]

Vol. 399: Edwards, C.; Lombaerts, T.; Smaili, H. (Eds.):
Fault Tolerant Flight Control
appro. 350 p. 2010 [978-3-642-11689-6]

Vol. 398: Hara, S.; Ohta, Y.; Willems, J.C.; Hisaya, F. (Eds.):
Perspectives in Mathematical System Theory, Control, and Signal Processing
appro. 370 p. 2010 [978-3-540-93917-7]

Vol. 397: Yang, H.; Jiang, B.;
Cocquempot, V.:
Fault Tolerant Control Design for
Hybrid Systems
191 p. 2010 [978-3-642-10680-4]

Vol. 396: Kozlowski, K. (Ed.):
Robot Motion and Control 2009
475 p. 2009 [978-1-84882-984-8]

Vol. 395: Talebi, H.A.; Abdollahi, F.;
Patel, R.V.; Khorasani, K.:
Neural Network-Based State
Estimation of Nonlinear Systems
appro. 175 p. 2010 [978-1-4419-1437-8]

Vol. 394: Pipeleers, G.; Demeulenaere, B.;
Swevers, J.:
Optimal Linear Controller Design for
Periodic Inputs
177 p. 2009 [978-1-84882-974-9]

Vol. 393: Ghosh, B.K.; Martin, C.F.; Zhou, Y.:
Emergent Problems in Nonlinear
Systems and Control
285 p. 2009 [978-3-642-03626-2]

Vol. 392: Bandyopadhyay, B.;
Deepak, F.; Kim, K.-S.:
Sliding Mode Control Using Novel Sliding
Surfaces
137 p. 2009 [978-3-642-03447-3]

Vol. 391: Khaki-Sedigh, A.; Moaveni, B.:
Control Configuration Selection for
Multivariable Plants
232 p. 2009 [978-3-642-03192-2]

Vol. 390: Chesi, G.; Garulli, A.;
Tesi, A.; Vicino, A.:
Homogeneous Polynomial Forms for
Robustness Analysis of Uncertain Systems
197 p. 2009 [978-1-84882-780-6]

Vol. 389: Bru, R.; Romero-Vivó, S. (Eds.):
Positive Systems
398 p. 2009 [978-3-642-02893-9]

Vol. 388: Jacques Loiseau, J.; Michiels, W.;
Niculescu, S-I.; Sipahi, R. (Eds.):
Topics in Time Delay Systems
418 p. 2009 [978-3-642-02896-0]

Vol. 387: Xia, Y.; Fu, M.; Shi, P.:
Analysis and Synthesis of
Dynamical Systems with Time-Delays
283 p. 2009 [978-3-642-02695-9]

Vol. 386: Huang, D.; Nguang, S.K.:
Robust Control for Uncertain
Networked Control Systems with
Random Delays
159 p. 2009 [978-1-84882-677-9]

Vol. 385: Jungers, R.:
The Joint Spectral Radius
144 p. 2009 [978-3-540-95979-3]

Vol. 384: Magni, L.; Raimondo, D.M.;
Allgöwer, F. (Eds.):
Nonlinear Model Predictive Control
572 p. 2009 [978-3-642-01093-4]

Vol. 383: Sobhani-Tehrani E.;
Khorasani K.;
Fault Diagnosis of Nonlinear Systems
Using a Hybrid Approach
360 p. 2009 [978-0-387-92906-4]

Vol. 382: Bartoszewicz A.;
Nowacka-Leverton A.;
Time-Varying Sliding Modes for Second
and Third Order Systems
192 p. 2009 [978-3-540-92216-2]

Vol. 381: Hirsch M.J.; Commander C.W.;
Pardalos P.M.; Murphey R. (Eds.)
Optimization and Cooperative Control Strategies:
Proceedings of the 8th International Conference
on Cooperative Control and Optimization
459 p. 2009 [978-3-540-88062-2]

Vol. 380: Basin M.
New Trends in Optimal Filtering and Control for
Polynomial and Time-Delay Systems
206 p. 2008 [978-3-540-70802-5]

Vol. 379: Mellodge P.; Kachroo P.;
Model Abstraction in Dynamical Systems:
Application to Mobile Robot Control
116 p. 2008 [978-3-540-70792-9]

Vol. 378: Femat R.; Solis-Perales G.;
Robust Synchronization of Chaotic Systems
Via Feedback
199 p. 2008 [978-3-540-69306-2]

Vol. 377: Patan K.
Artificial Neural Networks for
the Modelling and Fault
Diagnosis of Technical Processes
206 p. 2008 [978-3-540-79871-2]

Vol. 376: Hasegawa Y.
Approximate and Noisy Realization of
Discrete-Time Dynamical Systems
245 p. 2008 [978-3-540-79433-2]

Vol. 375: Bartolini G.;
Fridman L.; Pisano A.; Usai E. (Eds.)
Modern Sliding Mode Control Theory
465 p. 2008 [978-3-540-79015-0]

Vol. 374: Huang B.; Kadali R.
Dynamic Modeling, Predictive Control
and Performance Monitoring
240 p. 2008 [978-1-84800-232-6]

GPSR Compliance

The European Union's (EU) General Product Safety Regulation (GPSR) is a set of rules that requires consumer products to be safe and our obligations to ensure this.

If you have any concerns about our products, you can contact us on ProductSafety@springernature.com

In case Publisher is established outside the EU, the EU authorized representative is:

Springer Nature Customer Service Center GmbH
Europaplatz 3
69115 Heidelberg, Germany

Batch number: 10584805

Printed by Printforce, the Netherlands